ENOVALE™

How to Unlock Sustained Innovation Project Success

ENOVALE™

How to Unlock Sustained Innovation Project Success

Dr. Greg McLaughlin • Dr. Vinny Caraballo

CRC Press is an imprint of the
Taylor & Francis Group, an **informa** business
A PRODUCTIVITY PRESS BOOK

CRC Press
Taylor & Francis Group
6000 Broken Sound Parkway NW, Suite 300
Boca Raton, FL 33487-2742

Printed on acid-free paper
Version Date: 20130812

International Standard Book Number-13: 978-1-4665-9208-7 (Paperback)

Visit the Taylor & Francis Web site at
http://www.taylorandfrancis.com

and the CRC Press Web site at
http://www.crcpress.com

Contents

List of Figures

List of Tables

Acknowledgments

I acknowledge all those people who have contributed to my understanding and appreciation of innovation.

I especially thank my dear wife, Heidi, who has freely given her wisdom, advice, and counsel concerning this book. Without her patience and love, this book would not be possible. I also thank the Divine Trinity, Father, Son, and Holy Spirit, for their help, love, and assistance.

In addition, a special acknowledgment to William (Buzz) Kennedy for his suggestion that we provide information for nonprofits and to Dr. William McKibbin for his help in understanding the role of "big data."

Dr. Greg McLaughlin

This book is a combination of knowledge and experience that requires effort and sacrifice. It is made possible because I had the support of those individuals who sacrificed their time and resources. To make this book a reality, we required a global team of collaborators to help us reach different parts of the world. I dedicate this book to all the members of the original Project Impact team who leveraged their networks and provided their perspectives to make this work possible. I would like to thank Dr. Carmen Joham and Dr. Stephen Boyle of the University of South Australia, and Dr. Leonardo Pinedas of the Universidad del Rosario in Bogota, Colombia, for all their collaboration and help. I also dedicate this book to Ada, Ervin, Michelle, and my late parents, Ervin and Yolanda, for all they have done for me.

Dr. Vinny Caraballo

About the Authors

Dr. Greg McLaughlin possesses a unique talent for taking strategic visions and ideas and turning them into operational realities. He creates value in organizations through his ability to solve complex problems, recognize hidden or unexplained data patterns, and by creating practical, ready-to-implement solutions. Over the past 30 years, Dr. McLaughlin has developed a passion for innovation excellence resulting in the creation of the ENOVALE™ process and his recent books: *Chance or Choice: Unlocking Innovation Success* (2013), *Leading Latino Talent to Champion Innovation* (late 2013), and *Innovation and Healthcare* (mid 2014). His colleagues refer to him as a "Renaissance Man" given his diverse set of life and work experiences, including hurricane forecaster, author, Six Sigma/Lean Guru, published songwriter, and deacon.

Dr. Vinny Caraballo is an expert in sustained innovation success, with an emphasis on leveraging the innovative capabilities of a multicultural workforce. His belief that all innovation emanates from human beings led him to co-develop ENOVALE, a framework to identify and align innovation culture and embed it throughout an organization to make innovation a predictable and sustainable part of operations. Dr. Caraballo has led and trained teams on several continents and consulted with some of the world's premiere technology and professional services firms. Prior to entering the private sector, he served as a U.S. Army aviator. He is the coauthor of two other books: *Chance or Choice: Unlocking Innovation Success* (2013) and *Leading Latino Talent to Champion Innovation* (late 2013).

Chapter 1

Introduction

Sustained innovation success is a goal that many continue to strive for in their businesses or organizations. The question remains: How do we accomplish such a goal and operate in this reality? Some believe that innovation is a "numbers game"—keep trying until something works or succeeds. This is an expensive and risky proposition. With this strategy, should you continue if the odds of success are 1:10 or 1:20? Would another approach be more worthwhile? An innovation strategy that contributes one success after the other would be the optimal choice of all leaders and managers. The problem is that much of the literature on innovation speaks to the importance for companies and organizations to be innovative, but does not focus on the process of becoming innovative. The opportunity to offer your customers and users innovative technologies, products, and services should be a reality. "Intense customer focus leads the prototypical excellent company to be unusually sensitive to the environment and thus more able to adapt than its competitors" (Peters and Waterman, 1982, pp. 77–78). Yet there is very little information on how to initiate and sustain an innovation management program. Leaders and managers, which are the crux of sustained innovation success (Peters and Waterman, 1982), are struggling to find strategy that they can use that will deliver consistent innovation success.

The authors' first book, entitled *Chance or Choice: Unlocking Innovation Success* (McLaughlin and Caraballo, 2013), provided a proven management process, using the ENOVALE® methodology, for identifying innovation opportunities through validated outcomes. An outcome is an objective bounded (constrained) by its requirements. This second book takes the outcome and provides a method from project initiation to completion. Although a generic process was provided in *Chance or Choice*, this second book refines and solidifies that process.

The format or layout of this particular book begins with a short review of what innovation means and how it is transformative for products, processes, or services. Once we have reacquainted our readers with innovation, we will then discuss a series of strategies for each of the three means of innovation. These strategies will provide a systematic process for initiating and conducting an innovation project. It is assumed that these projects will be conducted in an environment conducive to innovation, meaning that the business organization has implemented the overall ENOVALE solutions process at the strategic level within the organization.

In *Chance or Choice* an example was discussed regarding the need for a car that can drive itself. We said this was a great idea but not feasible at the time of writing. In just a few months, the impossible became possible. In 2015, car manufacturers will begin introducing these cars, and a handful of states are adjusting their laws and requirements so that these vehicles may operate safely within their boundaries. In this case, the need and the technology joined to make this a reality. Innovation begins with one or more needs. Think about increased safety, lower insurance costs, less highway patrols. This great opportunity is becoming possible. Given that the car is robotically controlled, the entire field of robotics will explode. Soon, robots will do daily chores, care for the sick and elderly, maintain security, and free up our time for other activities. In 20 years, people will not know what we did without these machines. They will change everything as we know it. Machines do not require benefits, healthcare, vacation, and holidays. Local governments will assign many police and fire duties to these machines. The very nature of work will change, and some will be hurt as the innovation sweeps the world. This will create a new set of "needs" that will further drive innovation.

As with the first book, business examples familiar to many are used to describe and highlight our philosophy, strategic elements, and success criteria. Every business and organization will experience differences in implementing sustained innovation. For some it will come naturally; for others it will require a more rigid structure. Whether simple or complex, the necessity for accomplishing this goal outweighs any discomfort. The authors understand that implementing an innovation strategy for product, process, or service is different, and we address the differences when appropriate. What is interesting is the amount of similarity between the strategies and that certain steps, such as validation, are critical for all innovation efforts.

A word of warning to all of our readers: The strategies for project management and innovation success are dependent upon when and how the organization or business has implemented the ENOVALE solutions innovation process management. Without ENOVALE, there are no validated outcomes from which to build an innovation project. Skipping the ENOVALE solutions phase would add great risk to the success of the overall project, since success would be more of a "hit or miss." ENOVALE solutions deliver a validated outcome (an opportunity) to the team. The second phase is to take the outcome and transform it into a specific product, service, or technology.

Excellent project management skills, in an organization or business, are an important tool for innovation project success. Businesses and organizations that have a functioning project management application will find the strategies presented here to be very similar (at times) to their approach. The authors caution experienced project managers to understand that our strategies center only on innovation projects. The return on investment (ROI) will be realized with the additional profit and competitive advantage gained when offering customers and users this innovative approach. Given the success that businesses experience with ENOVALE, the natural tendency is to reorient all processes to this process. Since the strategies or tools were developed and tested for innovation, there is no proof that these will function in normal day-to-day circumstances. This was not the intention of this methodology. However, our clients keep insisting that this is not true and that these techniques and strategies work well for any project throughout the organization. We make no claim as to the usefulness or applicability to noninnovation projects.

We know from our 60+ years of experience that leaders and managers will want to skip steps, thinking that the time saved will reduce the overall cost and will have little effect on the outcome. Skipping steps is a recipe for failure. Employing a project management strategy not specifically designed for innovation guarantees less than stellar results. Match the strategy to the desired outcome.

Understanding the connection between the individual and innovation highlights the need for well-formed practitioners. Managers need to select the best persons for the innovation implementation team. Rather than identifying a potential opportunity, at this point, the team begins the task of implementing the opportunity. Managers can easily use the team developed for identifying the opportunity, and then substitute members when those who are more knowledgeable of topics, implementation principles, and practices are available. Team membership is often fluid and changes due to many circumstances. The key task is alignment to the project objective.

This second book builds on the knowledge acquired during the ENOVALE solutions phase and brings these outcomes into the real world. The authors use ENOVALE again just for easy recall purposes. These implementation steps move the outcome to a rational conclusion—build (initiate) or reject. The steps in part 2 of the ENOVALE strategies focus on creating the item (product, service, process, change, or technology), ready for the customer or user.

What our readers should expect is to learn the five implementation strategies that you can apply to any type of innovation project. Each strategy has its own seven steps and is designed for a specific innovation theme of concept. Partial application of the strategy risks reduced opportunity, limited success, and limited sustainability.

Finally, as always, we look forward to receiving comments and questions from our readers. We use these to refine and hone our strategies and topics. We continue to work toward a more automated system, one requiring much less labor. However, the fact that innovation begins and ends with the individual does not negate the overwhelming importance of people. Our task is to guide the process to a successful and sustained conclusion.

Summary

This book is the next rational choice in unlocking innovation success. Its sole purpose is to provide a credible stepwise process to implement a validated outcome. If that term is unfamiliar, then the authors suggest reading *Chance or Choice: Unlocking Innovation Success*; if it is familiar, then apply each themed strategy to a specific innovation project.

Chapter 2

Innovation Primer

The problem lies in the fact that innovation, as a concept, entails many different definitions and meanings. Researching the word in the dictionary, innovation generally refers to two keywords: *new* and *novel*. Accordingly, the definition is that an innovation is a new idea or novel (unique) concept. The definition limits innovation to only those products, services, or technology that is new. The question is whether an innovation can be something more than an invention.

As with all forms of innovation, inventions transform the way we do things; they shift the boundaries of our knowledge. The U.S. Patent Office receives numerous applications each day to protect inventions, but to most people these innovations are unique and special—not something thought of as ordinary or repeatable. Inventions occur irregularly, but when they do appear, they are generally described as groundbreaking. They are truly special events—not often repeated or replicated. For many, the "holy grail" for continuous innovation is a strategy that imitates and sustains innovation on a regular basis. Yet, inventions are too few to sustain a regular innovation effort. If an innovation is an event, what other types of events or activities can generate the same reactions as those only associated with an invention? Since innovation begins and ends with the individual, should not the definition include how the individual perceives innovation—what it will do for him or her?

Although an idea can initiate innovation, it must be in tandem with a need. Idea generation stimulates the thinking process. Innovation places ideas and needs into action. This is why we question the process of collecting only ideas from employees. There is a possibility of finding a "gem" of an opportunity (over the long haul the odds are about 10%), but that leaves a 90% chance of finding nothing worthwhile. Innovation becomes a numbers game—creating a large number of ideas until the "right one" is found. While we welcome ideas, we know that innovation initiates when a need persists. For those organizations with an idea entry process, modify the requirements to capture the need and the idea. If the need is not compelling or possible, then catalog the idea until such time both the idea and need align.

Therefore, innovation requires a broader definition or one focused on what the innovation will accomplish. That strategy would not restrict innovations to only "new" outcomes or ideas, but also include innovations that improve or change products, services, and technology. In fact, the term could easily apply to people and decisions that have resulted in a positive outcome. By focusing on what innovations can produce, we can better define how innovation transforms products, services, processes, and technology.

One recent article (Baregheh, Rowley, and Sambrook, 2009) reported 60 individual definitions of the word *innovation*. Obviously, innovation means many things to different people. Some see innovation as a novel idea, a new and unique product, or new technology. Innovation occurs when humans employ a creative process to meet a particular need: innovation begins at a very human level. You could even call this the "organic" level. Therefore, a correct definition should include how individuals view and judge innovation. Humans address a need, and how that need becomes a reality is the process of innovation.

From our studies and empirical research with Project Impact, we arrived at a definition that describes innovation from a perspective of what the innovation will accomplish. This will help to better clarify how individuals perceive (understand) innovation. We perceive innovation based on our knowledge and experiences. We use this knowledge and experience we have with the product, service, or technology (which we refer to as an item) and its performance to judge whether it is innovative. When the item performs better than expected and it meets a new or existing need, it is innovative.

To better clarify a definition of innovation that will relate to how an individual perceives innovation, Baregheh, Rowley, and Sambrook (2009) decided to examine the "means of innovation" (p. 1334), that is, understanding how innovation "transforms ideas into new, improved or changed" (p. 1334) outcomes, services, or people. From this definition and our research, we arrived at three unique descriptors of innovation that transforms objects (things) to meet new or different needs.

The three main descriptors or themes are new, improved, and change that describe how the product, technology, or service is transformed; in other words, how the product, service, or technology is "transformed" into something we define as innovative, as it better meets (satisfies) our needs. There is a distinctive strategy for each of the three descriptors. When a customer or user experiences a product, service, or technology designed to exceed more than what he or she expects, then we identify this as innovation. As expectations change, so do our needs change, and what was innovative becomes routine over time.

Individuals are the best judge of whether something is innovative or not. Where we differ from most authors, scholars, or researchers is that innovation is more than something new; it is more than a creative idea or new technology. It is a means to meet a need (new or existing) with something better than presently exists.

Think about digital photography—is this innovative? Well, if we use the standard terms of innovation, is it new technology or new to the marketplace? Is it a repackage of traditional photography or something very new? In fact, digital photography uses a very different mechanism to create a photograph. It is truly innovative!

We define *innovation* from its ability to meet a human need. The fact is that digital photography is innovative because it is new and meets a different or yet unmet need. The need was for a filmless system with instant development. The perceived savings in film costs was remarkable. However, a digital image does not enable the user to hold or display the photograph. Of course, you can keep the images electronically stored, but this is less personal. People still want the photograph, so photo processing remains a viable business.

Consider a second piece of technology, the cell phone. Is the cell phone truly new or novel or an improvement to existing technology? Given that the cell phone became popular beginning in 1990, is the cell phone today truly something new or unique, or rather quite an improvement over its 1990 predecessor? We believe that most people will in fact say that their cell phone today is an improvement over what existed previously. That being the case, the cell phone of today is innovative, as it fulfills many needs (that were unsatisfied in the 1990s) compared to its 1990 predecessor. Innovation occurs numerous times when improvements or changes occur to an existing technology, product, or service, given a changing (expanding) set of needs.

Defining the Means of Innovation

As mentioned previously, innovation consists of three distinct themes or concepts. Each concept or theme describes the transformation of a product, service, or technology into something perceived as innovative. The process of transformation is what a producer, manufacturer, or designer does, and the user (customer) then judges it as either innovative or not. After transforming the product, service, or technology, the user (customer) can classify (identify) the means of innovation as new, improved, or changed. Think of these as three distinct transformations of accomplishing the same goal, innovation. For example, consider that stocks, bonds, or commodities are methods (means) to invest money. All accomplish the same goal (hopefully), classified by type of investment. Each investment type requires a unique (and often related) strategy, even though the end goal is the same. Innovation follows the same pattern, as it satisfies new or existing needs by offering something better than its predecessor offers.

There are three means (or ways) of innovating. We refer to these means as themes:

Theme 1: *New* (something new or a novel (unique) idea)—normally we think of an invention.

Theme 2: *Improvement* (making something better)—this relates to products, processes, or services. Performance measures the amount of improvement. For those products, processes, or services that are underperforming or those for which increasing performance would yield additional competitive advantage.

Theme 3: *Change* (replacing what presently exists for something different)—affects people both physically and emotionally. Innovative change is positive change benefitting the individual and the organization (McLaughlin, 2012).

Each theme is distinctive, yet interrelated, as all need the individual to initiate the innovation. What is different is how individuals perceive their importance. This book presents a strategy, based on the ENOVALE® methodology, for successfully implementing new, improved, or change innovations. Of all three themes, innovative change is the least acknowledged. This book will provide a strategy for accomplishing innovative change in any business or organization. Making change positive is truly innovative, as it directly affects us all. Surprisingly, people frequently recognize it as being the most important and significant theme (descriptor) of innovation. Think back when change occurred in your job, your boss, or your life. Was it positive? Did it make a difference? Did it change you for the better? If it did, it was innovative. If the eventual outcome was positive, then the change was innovative. Of course, a negative experience may have also affected you. Negative experiences can easily lead to a negative outcome that can define a destructive change. Negative experiences can also result from destructive change—change made with a negative intent. Many times the situation warrants a negative outcome, but the process, the communications, and the repercussions can lead to destructive change. Destructive changes affect morale, motivation, and productivity issues. These are in fact anti-innovative, and the consequences and repercussions may be devastating.

Innovation transforms products, services, and technology into a reality that individuals define as innovative. When we define *innovation*, we use these terms not as a definition, but more about expectations (related to experiences) and perceptions (how we judge performance). Remember, innovation begins and ends with the individual. If we experience something much better than expected and it meets more of our needs, then we say it is innovative. Without getting into detailed specifics, we have scientifically validated these three themes (means) of innovation with numerous cultural and ethnic groups worldwide. More information on our research is contained in *Chance or Choice: Unlocking Innovation Success.*

There is nothing more compelling than asking individuals to define *innovation* and listening to the various definitions that both complement and conflict with one another. The divergence of opinion validates our premise that a definition also is insufficient. Given the diversity of opinion, it is easy to understand how conflict, dissent, and disengagement cripple the chances for success. Rather than defining *innovation*, it is best to understand how it is transformative and how the outcomes identify with one of three themes.

Perceiving Innovation

We all perceive innovation in our own ways. We use knowledge and experience of products, services, and technologies to determine how to satisfy needs and if we designate the outcome as innovative. Businesses and organizations need to be aware of what the customer or user requires, combined with their available resources and company objectives. Obviously, from a business perspective, we cannot know what each person perceives as innovative. Leaders and managers need to determine whether a product, service, or technology requires a transformation in order for a majority of people to recognize it as innovative. The amount of transformation will depend upon the need and expected performance.

Figure 2.1 describes the Global Targeting innovation outcome model from a producer/provider and customer/user expectation format. This is a minor revision from the model presented in our first book. The business or organization must

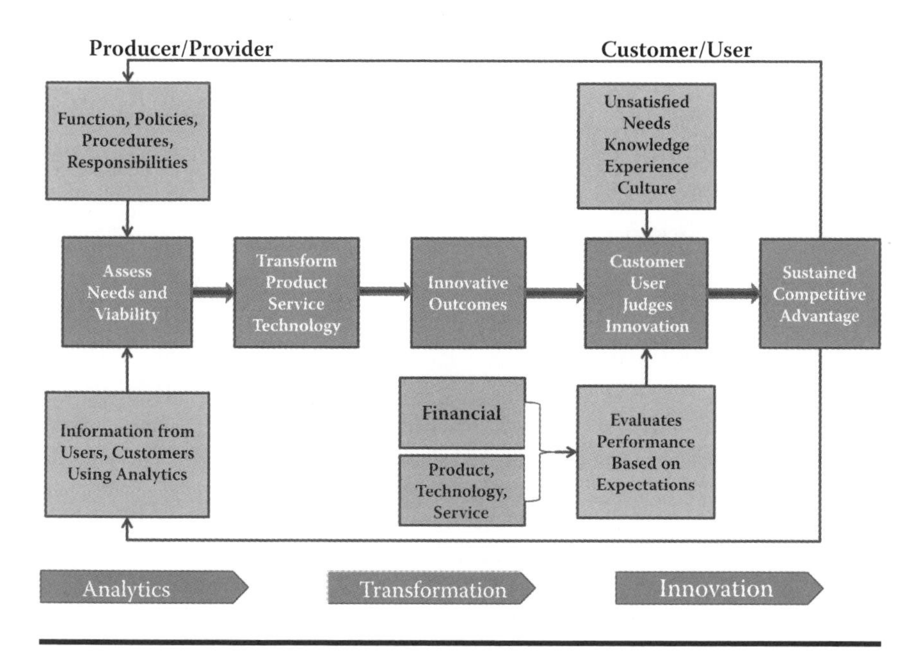

Figure 2.1 Innovation outcome model (2013).

research and categorize the needs of the customer or user and balance these with the resources available. Information gathering is a key to meeting unsatisfied needs. Analytics becomes a critical lynchpin to filter and interpret the data to extract usable information. This information gives the organization a distinct competitive advantage. Customers will then judge the outcome with their experiences and knowledge. Combined with an assessment of performance, they then judge the outcome as either innovative or not. The process of transforming the outcome is the result of the ENOVALE innovation process.

Transformations can take many forms. It could be a very new technology, a vastly improved product, or a significant change to personnel. The process of transforming an outcome is required for an understanding of its requirements (both functional and user), the objective it will accomplish, and what the item will become. It is not enough to "make something better"; it requires the product, technology, process, or service to meet more than existing needs. Customers respond (by purchasing) when producers better satisfy their needs. Meeting new or refining existing needs can accomplish the same objective. The trick is finding the balance between meeting needs and determining how much transformation is required. By knowing what and when to transform, businesses or organizations can achieve real competitive advantage.

For new products, services, or technologies new knowledge is often required. Many businesses begin with a small group of innovators who familiarize customers/users with the new or improved product. A perfect example of this is the Apple iPad, which essentially replaced the tablet PC. What made the iPad innovative was that it met additional needs. For the Apple iPad it was a small group, the so-called pioneers, that purchased the iPad. They identified needs such as size, weight, function, performance, and application that far exceeded existing tablet computers. The iPad was proclaimed as innovative and sales rapidly increased. The pioneers transformed perceptions of what a product such as the iPad could accomplish, and specifically how it exceeded those needs. Many times communications, marketing, and advertising favorably transform perceptions. However, unless additional needs are satisfied and performance exceeds expectations, it will not be perceived as innovative. Transforming perceptions is not enough, as the physical product, process, technology, or service must perform.

Innovation is not just about meeting or exceeding needs; it is also about meeting or exceeding performance. At times, we will find that people will identify innovation as purely performance driven, as the object meets all critical needs. We also know that individuals identify improvement when needs are met in a realistic time frame. It is best to understand that people perceive something as innovative because it exceeds the performance of something with which they have experience. Again, for the businessperson, is not important to try to make everyone recognize that what you offer, what you sell, is innovative. What is more important is to understand that the organization should be addressing what a customer or user needs, and how the product, service, or technology that it produces can deliver better performance.

Summary

Innovation begins and ends with the individual. Needs drive innovation; a growing population will require additional needs, thereby requiring more innovation. Of course, meeting all needs is impossible. Innovation brings about opportunity to businesses and organizations. Those who meet needs (those of today and in the future) with innovation will have sustained success and competitive advantage. Those who innovate infrequently will achieve some success, but at the mercy of their competition. The path is open to those who innovate and those who do not. Innovators will continue to add value to society and reap the benefits of their labor.

Chapter 3

Innovation Strategy

The *Merriam-Webster Ninth New Collegiate Dictionary* defines *strategy* as (1) "the science and art of employing the political, economic, psychological, and military forces of a nation or group of nations to afford the maximum support for adopted policies in peace or war"; (2) "a careful plan or method; the art of devising or employing plans toward a goal" (2000, p. 1165). A strategy involves science, art, plans, methods, and goals! It is a human process to devise a method to achieve a goal. The goal is successful innovation on a continuous basis within the organization.

In this chapter, we will dissect this description as it relates to innovation. A true misconception exists regarding this word and its meaning. Some see strategy as a series of precise, well-calculated moves to accomplish an objective. Others see strategy as a process to obtain an objective, and still others see strategy as more of a mechanism for driving an objective. That is, one strategy is precise, well-thought-out moves (as in playing chess); another is the method used to achieve an objective (win a war); and finally, the third is more of a managerial directive that provides a specific set of activities to be accomplished. *Strategy* is a word well used, with multiple, sometimes overlapping meanings.

Strategies from the perspective of Global Targeting involve the framework needed to implement an innovation project. Each step builds upon the former; the information gathered is critical for evaluating the innovation. Information gained from the customer or user provides necessary data on items such as needs, purchase behaviors, consumer expectations or perceptions, and competitive advantage. The strategy becomes the plan to achieve the objective. Considering the strategy only as a plan is limiting, since the strategy contains the information, experiences, and validation of previous steps to succeed.

Limiting yourself to one mode or a very narrow definition of *strategy* limits its overall impact. You could begin with Sun Tzu's *Art of War*, but we are not planning for a battle, but rather looking for an opportunity. This is a key point for innovation—it is an opportunity that will be the focus. Strategies for finding an innovation project, with a high opportunity potential, were discussed in our *Chance or Choice: Unlocking Innovation Success* (2013) book. This book describes how to create, evaluate, and measure performance of an innovative outcome.

Outcomes meet requirements (user and functional), have limited risk, and are validated for performance. Outcomes are opportunities waiting for implementation. The opportunity, if accepted, becomes the innovation project. The ENOVALE® solutions process creates outcomes; outcomes initiate the innovation project. A decision to proceed on an opportunity comes from a proven outcome, rather than purely on judgment or opinion. If the outcome performs, and its viability is proven (the opportunity identified), then the decision to proceed (or terminate) can commence. Making the final decision requires judgment, experience, and opinion. That is, evaluate the outcome from a high-level strategic perspective: Does it meet company or organizational objectives? Do the resources exist? Are costs outweighed by the benefits? We know from experience that ENOVALE steps yield much information about the outcome and should provide 80% of the information needed to make a final decision. To date, every outcome that was accepted proceeded to the project stage.

Given the three themes or means of innovation, Chapters 4–6 develop a unique strategy for each innovation type. There are three strategies for new and unique forms of innovation: two for improvement and one for change management. Each strategy used the acronym ENOVALE.

By modifying the ENOVALE process and identifying the type of innovation desired, we will present a strategy for each project type.

Strategies for Success

The strategy begins with a decision related to implementation of the outcome. Evaluate the strategy based on three criteria:

1. The needs of the customer
2. The performance expectations of the outcome
3. Operational concerns (cost, benefit, ROI, competitive advantage, etc.)

If the customer needs a unique, new product, service, or technology, then the strategy developed for distinctly original items is recommended. These strategies include a separate methodology for distinctively unique items, items requiring a new feature or new use, and items requiring a new approach. Inventions are distinctively new items. A new feature of use provides additional life cycle support. A new approach is one undertaken by marketing and advertising to "breathe new life" into an existing product, technology, or service. For the criteria above, the needs outshine the remaining criteria since little is known of operational or performance expectations.

If customer or user needs require improvement in an existing product, technology, or service, then choosing the improve strategy would prove beneficial. Improvement subdivides into two categories: items that are underperforming and those that require better performance to increase purchase behavior and competitive advantage. For those items that are underperforming, a specific strategy to determine root cause reasons and probable solutions is implemented.

Finally, for an opportunity to change, an existing process, product, service, or process requires a strategy that drives a positive outcome. For this strategy, the decision-making and implementation processes are critical to the results. This strategy examines the consequences of change and its effect on resources, processes, and employees. Innovative change is a reality when the experience of the outcome is positive.

Although there is some cross-integration of certain terms, each desired outcome drives its own strategy. Mixing and matching are not recommended and should not be an option. An improvement strategy would be meaningless if the customer/user desires a "new" product, technology, process, or service. What is clear is that the outcome drives the strategy selected.

As with all strategies, there will be a need to make some modifications, adjust or modify the theory, and adjust for changing business needs. Strategies are dynamic, not static. As the nature of work changes, so will the strategies that define this work.

Summary

In summary, this book presents implementation (project) strategies for the three distinct themes of innovation. In all cases, the term *ENOVALE* is used, but with distinctly different meanings. Continuing the process, established in the first book, there is a strong emphasis on satisfying needs and meeting performance requirements.

Chapter 4

New Technologies

Introduction

Most people think of innovation as that associated with something that is unique, original, or new, commonly called an invention, and for the most part, these are technological advances, although innovations that are new also come in terms of products, services, and processes. For Global Targeting, when discussing innovations that are "new," the term refers to the originality of the innovation. In this case, original means something unique, something that has found a new application, or even a new approach. For this section and this chapter, the focus will be on original innovations. That can include technology, product, ideas, concepts, or new ways or methods of accomplishing a goal or objective.

ENOVALE® begins with those innovations classified as truly unique and original. The keyword for this strategy is *unique*. Global Targeting has developed a strategy for implementing unique and original items. The second approach identifies a new application or use for an existing item. This is a common strategy for items with long life cycles. Consider the light-emitting diode (LED)—first used in equipment, then for lighting, and now for Christmas tree lights. There seems to be a new application found every year. Finally, finding a new approach is also an aspect of originality. For example, an item such as laundry detergent essentially is the same item it was when first created. The only difference has been in the packaging, formulation, and additives applied, used to modify the product. Therefore, the product remains the same, but the provider or producer has added a new approach to that item by modifying it slightly and calling it "new and approved." The wording is familiar to most, but in fact, the product is not truly new or original; the approach is new.

Unfulfilled Needs

Before beginning the ENOVALE strategies process for truly original designs, first consider any unsatisfied needs. These are needs that the customer or user has said that he or she would prefer, want, or desire if the technology, availability, and cost were within his or her range of acceptance. These unsatisfied needs provide an excellent opportunity to develop truly original items with the potential of having a huge impact on the customer or user. It is a need that has not yet been met or satisfied. This counteracts the idea that people do not know what they want or desire. There will always be incidences when an invention comes directly from an idea that finds a need. Businesses must create a process of collecting and evaluating needs. These unsatisfied needs define future opportunities.

Think about even the most revolutionary new items that have arrived to the marketplace in the last 20 years. When you think about an item, think about the need that preceded the item. Was the need obvious? Was it on someone's wish list? On the other hand, did the item appear without satisfying an obvious need? For Global Targeting clients and customers, satisfying unfulfilled needs will move businesses ahead of the competition.

In order to collect information on unfulfilled needs, the organization requires a mechanism for collecting and sorting the data. One traditional method is through focus groups, surveys, and customer assessment. The second, more radical approach is with the use of "big data." Big data (sometimes called the fourth paradigm) (Hey, Tansley, and Tolle, 2009) are the technology and practices of handling voluminous amounts of structured and unstructured data (Brust, 2012) that require specialized analysis. This provides up-to-date information on unsatisfied or previously unknown needs. Remember not to confuse the need with an idea. Idea generation is extremely important in conceptualizing the need. In fact, many organizations miss the need in favor of generating ideas, thereby often reducing the overall effectiveness of the innovation. For those companies and organizations that collect ideas or create sessions to generate ideas, be aware that the idea is not going to necessarily identify a potential innovation project. Global Targeting recommends that organizations collect ideas with corresponding needs. This will help greatly in evaluating potential candidates, at this early stage, for implementation.

Many organizations trust individuals in research and development departments as well as the engineering department to create opportunities for innovation. This can be a successful strategy for organizations quite familiar with innovation. It can also be unsuccessful because these groups do not consider needs as the generation or starting point of an innovation. That is not to say that these organizations are not true innovators, but it does say that relying on only these departments is dangerous for the organization. This is especially true in industries such as process and technology industries that rely heavily on new products or new applications for products or new approaches. These organizations must realign their efforts to get ideas and needs in pairs.

Most organizations do not have a separate research and development group, or even an engineering group, and therefore their innovations must come from their customers or users, or internally from employees and managers. Here the refocus on needs may be easier since the organization is not bound to a preconceived method to generate innovation opportunities.

An unsatisfied need is an opportunity that has not yet been created or initiated. One simple example is high-speed air travel. The Concorde served a limited number of airports but delivered speeds twice that of sound. Even though the Concorde is not available today, the need to reach a destination quickly is still a priority. This remains, then, an unsatisfied need. The assumption is that passengers would pay for such a convenience, if it were available.

The final question is whether needs and ideas share a common beginning. Are needs just ideas that have found acceptance? The answer is a resounding no. Needs describe a basic requirement for survival (basic) and growth (motivational and achievement). An idea remains a thought until fully implemented. This is not to diminish the fact that an idea may generate a connection to an unsatisfied need. Ideas can also help clarify or satisfy aspects of these needs. The fact, however, is that the need drives the innovation. Ideas should provide the initial framework for innovation, yet over time, they can become less and less effective. One organization noted that ideas generated 40% of the innovations in the first year, but by the third year generated only 10% of the innovations. Dahl, Lawrence, and Pierce (2011) tell us that ideas are not the place to begin to consider a potential innovation, but ideas provide information that assists in defining needs. Generating ideas is extremely useful and part of the ENOVALE process, but the idea must coincide with an unsatisfied need.

To identify unfulfilled needs, ask the following questions:

1. What do customers want but cannot yet see the connection to your business?
2. What is the nature of their complaints and suggestions for improvement?
3. What is their predisposition to you as a provider if these unfulfilled needs are not satisfied?

4. Think about the internal needs of the organization. What would dramatically improve competitiveness and profitability of the organization if a particular need or set of needs were satisfied?
5. Now, think outside the box. What's not supposed to be able to be accomplished, but if it were it would greatly change the nature of the business?
6. What customer needs remain unmet today?

Identify the opportunities that arise from this set of questions and use them to initiate a need that could lead to an innovation.

After identifying the unique need, begin the ENOVALE solutions process. Do not begin at the ENOVALE strategies process (which takes a validated outcome or concept to reality) without first completing the seven steps of ENOVALE solutions. Once the outcome is validated, the need is to move the original innovation to the ENOVALE strategies phase (Figure 4.1).

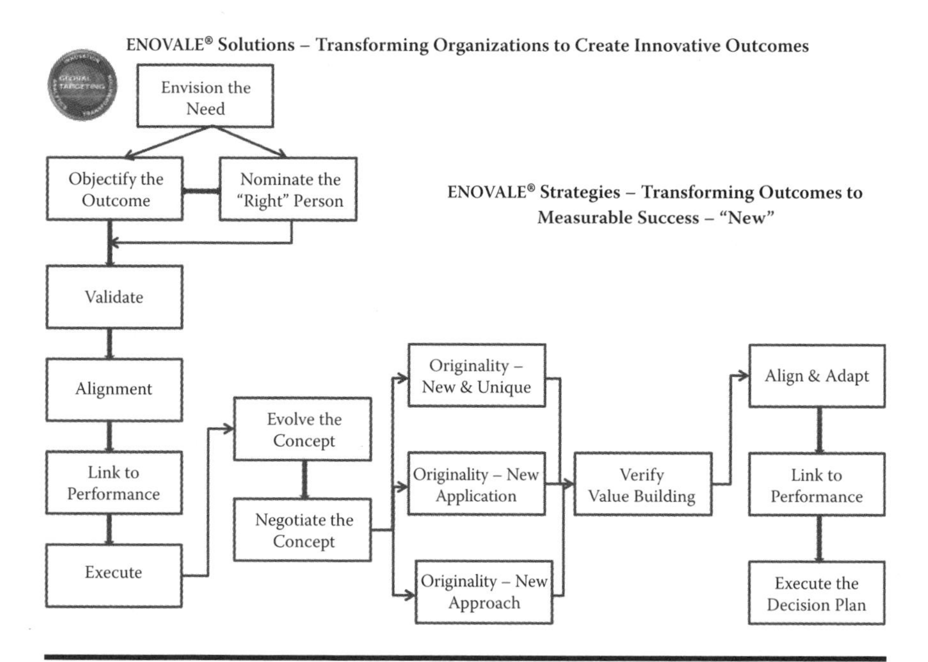

Figure 4.1 ENOVALE solutions and strategies—new.

E—Evolve the Concept

Beginning the process of implementing a project requires a valid outcome, established during the ENOVALE solutions phase. Just as a quick refresher, outcomes are objectives bounded by their requirements. Outcomes address limitations and assumptions. In this sense, outcomes are well-refined and validated objectives, aligned with the team working on the project, and linked to specific organizational performance measures. Although further refinement of requirements, limitations, and assumptions is always required, outcomes are ready for implementation. This section will take the outcome to its conceptual stage (Figure 4.2).

The validated outcome contains a wealth of valuable information. Outcomes reveal how the item will perform, how it will function, and what limitations to expect. Additional information reveals limitations and assumptions. The outcome provides the focal point for the innovation project. Yet, the outcome is only an objective—a well-developed and evaluated

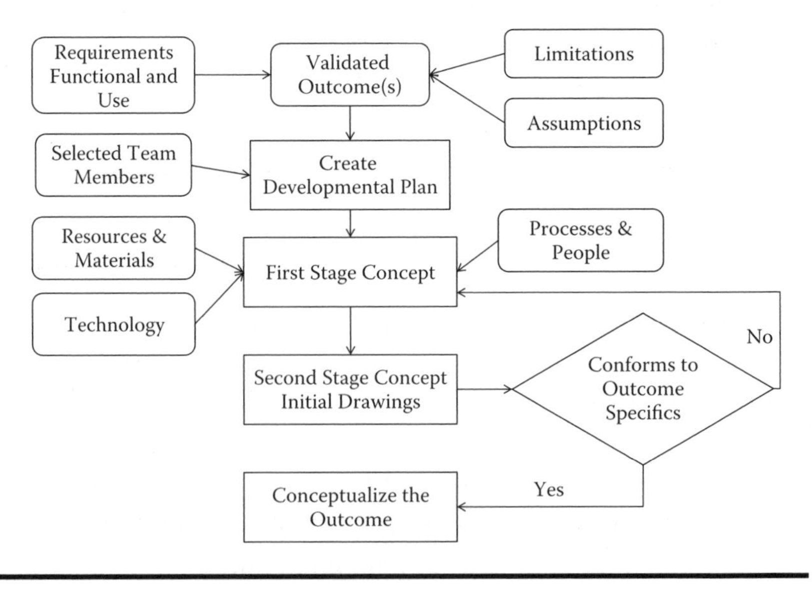

Figure 4.2 E—evolve the concept.

goal. The steps involved in the ENOVALE strategies process take the outcome and transform it into a concept.

To begin the transformation from outcome to concept, the project team needs to develop a plan detailing the process of transformation. The fundamental elements of the transformation are as follows:

1. Understanding and conceptualizing the requirements, both those that are functional and operational and those associated with the customer or user.
2. Understanding those limitations made to bring the outcome to reality. Limitations will certainly bring about unintended consequences. These consequences may be positive or negative, and for negative consequences, there will always be repercussions. Limitations define what is likely and possible with the outcome. Limitations also breed opportunities since what may be impossible today may be the new norm tomorrow.
3. Understanding the assumptions made regarding the outcome. Assumptions are risk based, since there is a need to deal with inconsistencies that exist within our environment. When making assumptions, there should be a corresponding understanding that certain consequences will occur when not meeting the assumptions. Assumptions, like limitations, change over time, requiring periodic evaluation. This is true as much during the conceptual phases as during the implementation and maintenance phases. The inability to control (predict) all assumptions is due to risk. Acknowledging risk and working toward a plan to minimize its effects is just good project management. However, the added uncertainty, associated with something so new, compounds this effect. Risk is an underlying deterrent to success. Evaluating risk requires an understanding of the possible consequences of failure.

The concept takes the outcome and transforms it into a physical form. That is, it makes the concept tangible. The tangibility of the concept is dependent upon the item's intended use. A service, for example, has both tangible and intangible elements, with a strong intangible measure of performance. Customers or users judge whether the service is performing or not. Their perception of service performance is in fact an intangible. Therefore, new items must consider both tangible and intangible elements related to an acceptance by the customer or user. This only reinforces the idea that at the concept stage, concerns about both tangibles and intangibles must be taken seriously.

The concept must address both tangible and intangible aspects of the item. For products and technology, it is much easier to grasp the tangible nature of the concept. No matter what the item, there is always an element of intangibility that exists and pertains to performance. For the concept phase, though, the focus is on the tangible elements of the item. The purpose? Bring the concept to reality! One additional concern is the process, that is, the process of converting outcome to concept, and concept to reality. This first stage transforms the outcome to become the first-stage concept.

Consider examples of outcome transformation. The elderly need an effective mechanism for dispensing prescribed drugs at home. The concept would entail a description of

- The delivery mode, how the concept controls and directs the core functions and monitors pill usage
- The time dispensed
- The resources and materials needed
- The people involved

The concept provides a description of an item's purpose and use. Whereas the outcome focused on identifying the item, objective, requirements, assumptions, and limitations, the concept uses that information and describes the process of taking the concept and making it a tangible. Converting the outcome to the concept must include a visualization or representation. For new innovations, this is converting the outcome, which is both unique and original, to reality. The concept is the outcome transformed into what will be a physical part, service, technology, process, or human activity. The simplest way to think about this is when someone applies for a patent, he or she must include a drawing or visualization of what the item will do. Essentially, this is what we describe as the concept. The difference is that it is not just a drawing or blueprint, but it also contains a treasure of information both from the outcome phase and from the development of the concept phase. To reiterate, inventing something does not necessarily involve innovation. Inventing something that's innovative requires that needs are satisfied and that the outcome is validated. The visualization may be a chart or graph; usually this is a drawing, which may or may not be computer aided, a blueprint, or a model. Just as a point of reference, mathematicians create equations to represent their concepts. That tells us the importance of visualizing the concept.

The development of the concept is a critical step in the overall developmental cycle. This can be a lengthy experience, spanning upon the complexity and distinctiveness of the outcome. Consider that the concept is essentially the information and visualization that an organization or individual will send to the patent office. More information, provided by the individual or organization, provides a better idea of the concept. It also protects the concept or idea. The concept must describe the item or invention and demonstrate how the innovation will perform. Individuals judge performance and decide with experience if a need is truly satisfied.

Summary

For the new theme or means of innovation, the outcome transforms itself into a concept. Moving the outcome to concept and then reality requires a development plan, a focused team (selection via the ENOVALE solutions process), and a strategy for implementation. The ENOVALE innovation strategy for new projects provides the mechanism for implementation. In this first stage (step), visualizing the concept into a set of drawings, specifications, or computer graphics is necessary for success. Visualizations help to conceptualize the outcome.

Discussion Questions

Assume that a new innovative outcome, as identified and validated through the ENOVALE solutions process, is ready to move into the next stage of development.

1. Describe how to transform this outcome into a concept. Remember to define a simple process that considers the wealth of information coming from the valuation phase of the ENOVALE solutions process.
   a. What must be known to move the project forward?
   b. How do requirements, assumptions, and limitations shape the concept?
   c. What "voices" (customer/user, R&D, operations, marketing/sales, etc.) must be heard?
   d. What objective does the organization or business expect?

N—Negotiate the Concept

After solidifying the concept, it is time to review the specifics. Why review the concept at this point? Simple—when deciding on a concept and bringing the concept to life via a visualization, it is time to revisit its basic premise and intent. Consider the following:

- Clarity and application

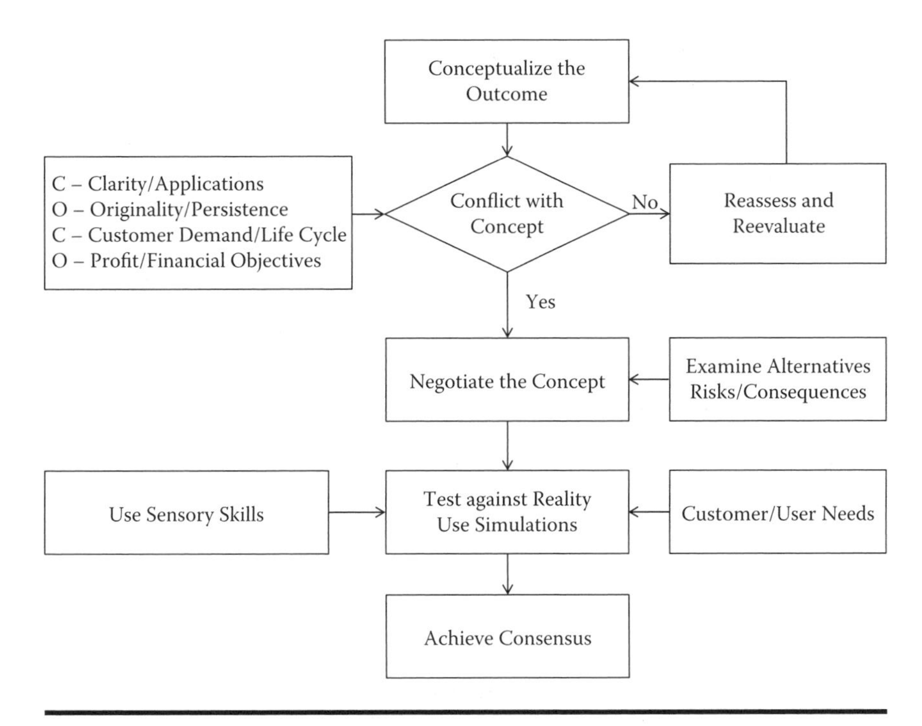

Figure 4.3 N—negotiate the concept.

- Originality and distinctiveness
- Customer demand and life cycle
- Objectives (profit and financial)

The COCO dimensions assist in clarifying the concept. Should conflict arise in reaching a concept, apply the COCO dimensions for comparison purposes. Use the matrix provided (Figure 4.3).

Below the COCO dimensions are defined:

Clarity: Full and complete understanding of how the concept will apply to perceived need.

Originality: Originality of concept; ability to stand apart and fend off competitors.

Customer demand: Meeting/exceeding the needs of the customer/user over the entire life cycle.

Objectives: Meets profit and financial objectives.

Table 4.1 COCO Concept Evaluations

COCO Criteria	*Concept 1*	*Concept 2*	*Concept 3*
Clarity			
Originality			
Customer demand			
Objectives			

The scale shown in Table 4.1 is generally descriptive (fully meets criteria, partially meets criteria, rarely meets criteria, for example). Other scales, created especially for the innovation, would be an improvement.

Clarity in intent and purpose is the key to success. If the objective is executable (doable) and worthy of implementation, then the concept passes its first test. Application refers to use and persistence to the time the item remains useful. The next great iPhone or iPad may be profitable, but at what cost? Innovations persist (long-time horizons) and are not easily copied or replicated. Life cycle here refers to the time the items remain competitive and profitable. Only the business or organization can seriously address this issue. Discussion, revision, and negotiation are the keywords of this phase. Better to make a change now (at this stage) than to attempt a change (modification or revision) at a later implementation stage.

Negotiation is both a science and an art. Negotiation involves more than one person and assumes both parties have viable points of view. Rather than subjecting the concept to a brainstorming and voting session, negotiation uses the experience, knowledge, and persuasiveness of the individual's negotiating skills. An approved outcome must transition to a concept and then reality. The project needs more than just a yea or nay vote. For those involved, it must be a win-win situation (remember alignment). When playing poker, it is not until your opponent shows his or her cards that the game ends (the negotiation is complete). For innovation, negation with hidden cards only slows and limits the process. All team members must freely exchange their ideas and negotiate their positions until the negotiation ends.

Global Targeting recommends negotiation since the decision maker should not be the sole person making the decision (this is probably the most difficult recommendation offered by ENOVALE). The science of negotiation is the offering of ideas/suggestions based on experience and judgment that affect bottom-line performance (if X changes, then Y changes). The art of negotiation is the skill of the individual to sell his or her idea/suggestion without evidence or validation. The key to negotiation is alternatives. There is always a better way or method to accomplish a task, decision, or a concept. Alternatives provide the wherewithal to discuss concepts derived from outcomes. Alternatives provide for a level playing field, and in addition:

- There may be a better method to accomplish the same outcome.
- Have a backup plan in case the original plan fails.
- Never assume that concept is an exact representation of the final innovation product, technology, process, or service.

Concepts will evolve to reality, and the process always exists in an open forum. The makeup for this negotiating team includes those individuals with a stake in the outcome. Those who do not share responsibility in the outcome's success or failure lack accountability. The team may decide to include (from time to time) individuals with expertise, but this occurs only when the team lacks the needed experience or knowledge.

Once negotiations are complete, begin the next step to reality. Producing a prototype, first/trial piece, or experimental unit is the next step in evolving the concept to reality. The purpose is to move from paper (concept) to a more visual and tangible item. For products and technology applications, a significant amount of time may elapse between concept and physical reality. This first piece of analysis provides for information on functionality and customer/user evaluation. When the concept is a model or simulation, apply a set of performance measures. Look for opportunities; be aware of consequences. The closer the business or organization can come to developing a physical item, the better the method of evaluation and comfort with the product. With items such as services, this is not possible except for the creation of a process flow diagram. Global Targeting recommends that service providers consider the use of models and simulations for best evaluation.

During this phase, evaluate risk and consequences. Examine life data, failure rates, breakdowns, and maintenance. When using models or simulations to test a service concept, resist the temptation to consider this a description of reality. As real as something may seem, it lacks the inherent variations (chaos) that make life interesting and worth living. Nothing replaces human knowledge and ability and the five senses to evaluate an item.

Finally, if a research and development (and engineering) department or group exists, do not assume these individuals should lead (or be responsible for) an innovation project. Innovation is organic; it begins and ends with the individual. These departments have experience and judgment (and the wherewithal) to conduct an innovation project. Their expertise makes them a valuable resource. Once negotiations are complete, these organizations may take the responsibility of developing the concept to full reality. Rather than demeaning their influence or expertise, these departments are now best prepared to complete the process of bringing the concept to reality. For those without these departments or functions, continue through the ENOVALE strategies process.

Summary

Negotiation is a key in developing new types of innovations. Realizing the benefit of negotiating the concept provides an open forum for all those to be aware of new opportunities while solidifying the concept. Negotiation uses the best skills of individuals to reach consensus and a win-win attitude. It is a key for organizational and project concept alignment. This is the final opportunity to evaluate and modify a concept before it becomes a reality.

Discussion Questions

What are the best traits for selecting someone to work on a new type of innovation project? Consider the following traits:

- Intuitiveness
- Creativeness
- Familiarity with methodology
- Tolerant of others' viewpoints and ideas
- Self-motivated
- Able to work in situations with few guidelines
- Focused on the project objective

Add others that may seem reasonable.

O—Originality

Once the concept is complete, the team can begin the process of conversion to reality. This is the step of maximum originality (creativity). The saying "the rubber meets the road" best describes this phase. Some would ask, is not creativity critical at the concept phase? Yes, it is, but this phase is where the item goes from concept to reality. This is where the driverless automobile goes from initial concept to working prototype. This phase determines how something will function and whether it meets its requirements. When NASA committed to President Kennedy's challenge to land on the moon, it developed the concept. Putting the concept into practice is truly the most difficult part of the process, and the one that requires the majority of originality and creativity. It is not that other processes do not require creativity (originality), but this is the phase where it is most critical. Truly original concepts require a greater need for creativity at this phase. Yet there are three new types of innovation:

- The need drives something totally new—invention (an item that has never existed).
- The need drives toward a new design with existing or modified components.
- The need drives a new approach for an existing product, service, or technology.

New (Original) Concepts

A very uncommon occurrence best describes this phase. Incidences of these occurrences are rare and generally unexpected. More often than not, most researchers consider these a failure. The fact that these items change the existing paradigm (rules) is often hard to accept. Consider the person in the nineteenth century who first saw his or her first photograph—the need always existed, but the means (technology) was not possible. It was truly an original innovation! Individuals would classify this as new. When first invented, the inventor did not understand the implications of his or her invention—how life changing it would become (Figure 4.4).

Unique characteristics of this type of innovation include the following:

- These new innovations can create new businesses (new industries) or new opportunities.
- These items appear when the need exists, given that technology and available resources are present. Certain types of inventions (discoveries) fall under this category:
  - Medical discoveries

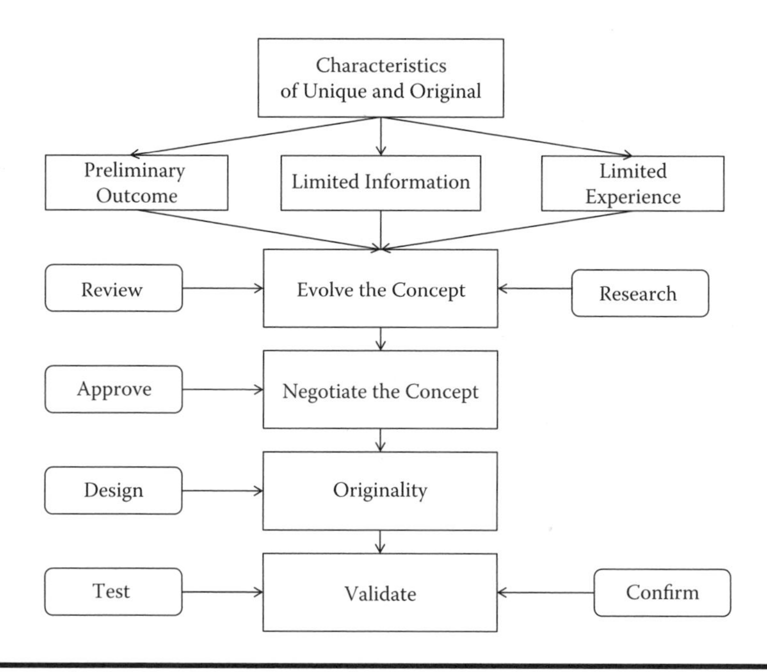

Figure 4.4 O—originality (new components).

- Defense-related technologies
- Internet businesses, such as social networking
- Electronics (for example, the microchip, transistor, etc.)

Originality requires a large amount of research and review. In addition, there is a need for secrecy and protection (of the intellectual capital). Even ideas can become intellectual capital and the proprietary capital of businesses and organizations. The strategy at this stage is simple: keep the efforts focused on the original. The elements of the strategy are

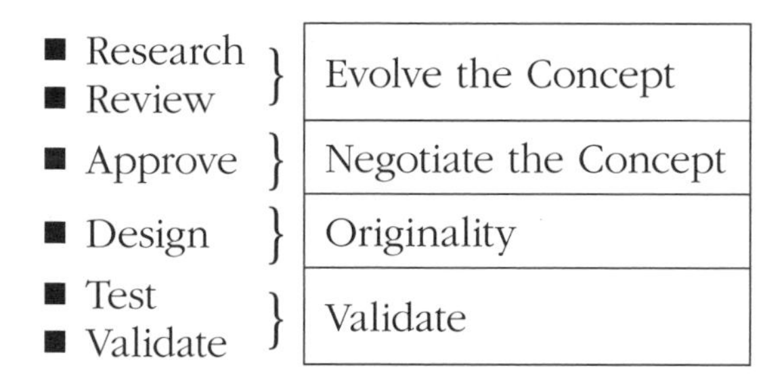

The steps are straightforward, but the strategy is conducted in secrecy to protect the intellectual capital. At each stage, there is a need for tightly controlled communications.

Summary

Truly new items require creativity, numerous resources, the right team, and a plan for execution. Given the uniqueness of many innovations, these plans change from item to item, adding costs and large time expenditures. Global Targeting proposes a more authentic approach using validated outcomes (from the ENOVALE solutions phase) as a method of initializing the innovation. The process saves time, energy, and resources. Confidentiality of information is secured and protected. Teams can focus on achieving the outcome. Moving the project from outcome to completion naturally involves R&D and engineering. The business or organization offers to the customer/user an item that meets specific needs. The more focused the needs, the longer it will take competition to offer a similar item.

New Design—Existing Components

During this stage, originality is linked to the final product or service. A new application or use (feature) may involve everything from delivery to appearance. What sets this apart is the fact that the innovation team will work with existing technologies and resources. This team will research the following:

- Potential product/patent infringements
- Product and technology research
- Customer behavior and psychology
- Existing and potential consumer demand

At this stage, the focus of originality is on what is new about the product, technology, service, or process. This new version of an existing product must meet additional or existing

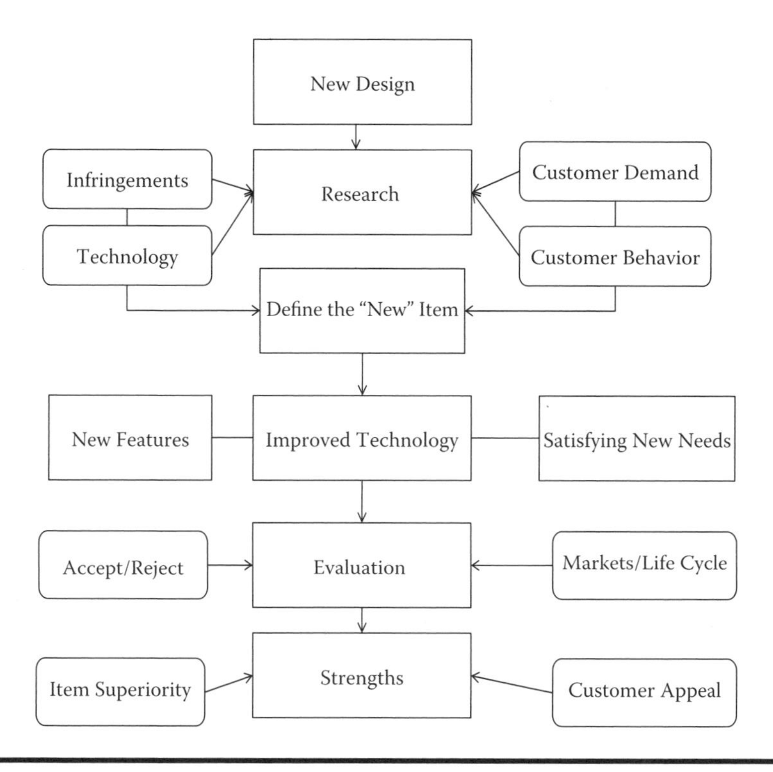

Figure 4.5 O—originality (new applications/uses).

needs by exceeding that performance which is presently available (Figure 4.5). Performance covers measures such as the following:

- Products—performance of the product per its requirements (specifications)
- Services—time to completion, effectiveness
- Technology—promised performance measuring capability

Performance is the measure of what a product, service, or technology provides. What the customer needs and receives is the prime driver. The iPad is an excellent example of this type of new design. The company uses existing resources and technology to offer the customer or user something beyond what he or she has experienced previously. This attitude is one where it is easy to achieve a self-deprecating environment.

This new aspect includes new features, improved technology, and a new application for an existing product or service. The primary objective is to convince consumers that the new use is a worthwhile purchase. To be successful, the organization must understand consumer purchase behaviors, customer/user expectations, and the meaning of *new*. Defining specific needs, understanding how customers evaluate performance is critical to success at this stage. This type of innovation is the most common type of new.

A new design with existing components is a simpler way to introduce users to your innovation. As with any new design, evaluation must take into account strong input from the public and practical applications. Examining the new design should consider the following:

- Consumer purchasing behaviors
- Technological product/service superiority
- Consumer appeal (ease of use, meets multiple or subordinate needs, resists duplication, exceeds normal (expected) life cycle)
- Costs (marketing, advertising, process) and benefits for the new application or use
- Specifics that distinguish new from previous versions

The most difficult aspect is to convince those that purchase that the new application, feature, or use is truly innovative. Innovation meets more needs by delivering better than expected performance. This phase (step) must deliver sufficient, relevant information that convinces the customer/user to experience the new aspects of the product, service, or technology.

Summary

A new application for a product, service, or technology is the one means of innovation with which everyone is familiar, from the large dosages of advertising communication we receive. For an item to be innovative, it must either exceed existing needs or meet new ones. When identifying a new use for a product, these needs must be the driving force. If scientists find a new application of an existing medicine, the first question must be, what needs are now not fully satisfied? For example, when working with GE Medical Devices, it was company policy in the 1990s to dispose of older equipment when it was replaced with a newer version. Working with individuals in Latin and South America, older devices (in perfect working order) such as computed tomography (CT) scanners were refurbished and sold at a discount to users in Central and South America. This way, recent and effective equipment could be provided where none had existed before.

New Approach

Finally, the third original approach is one where the new aspect is a new approach for an item within the scope of existing technology. Originality is limited to the item's new approach or new use. Finding a new approach is similar to extending the life cycle of the item. This new approach could have a very short life cycle before duplication or being superseded by existing competition. It does offer the approach of providing something new about an item without cost or time delays. A new approach or new feature might open up avenues of opportunity for an item. Probably the most recognized aspect is in product advertisements that claim that the product is new and improved. We know that it is impossible to have both new and improved innovation. However, if the customer finds that the new product, process, technology, or service meets a new need or further expands on an existing need, it may be deemed innovative. The fact that competition can easily replicate or duplicate these efforts limits the overall effectiveness of this particular type of design. It provides a wide range of information related to perceptions and attitudes (all necessary elements of customer purchase behavior). When advertising or marketing this concept, its unique application and new approach is communicated to the consumer. In simple terms, communicate to the user what is new, what needs are or will be satisfied, and how the item performs better than its predecessors do. Consumers will comprehend the benefits and, for those impressed by anything new, will understand this as innovative (Figure 4.6).

Clearly specify this new approach with its requirements clearly detailed. What distinguishes this new element from existing items is that it will fill unsatisfied needs. Therefore, rather than the focus being on design and development, it is on perception and attitude. Realizing that the consumer may be confused trying to determine what constitutes a new approach or use, it must be precisely communicated and the need satisfied. Differentiation is key. Rather than stressing only the new approach, stress the improved performance or specific needs satisfied. Consider the example of an IT resource

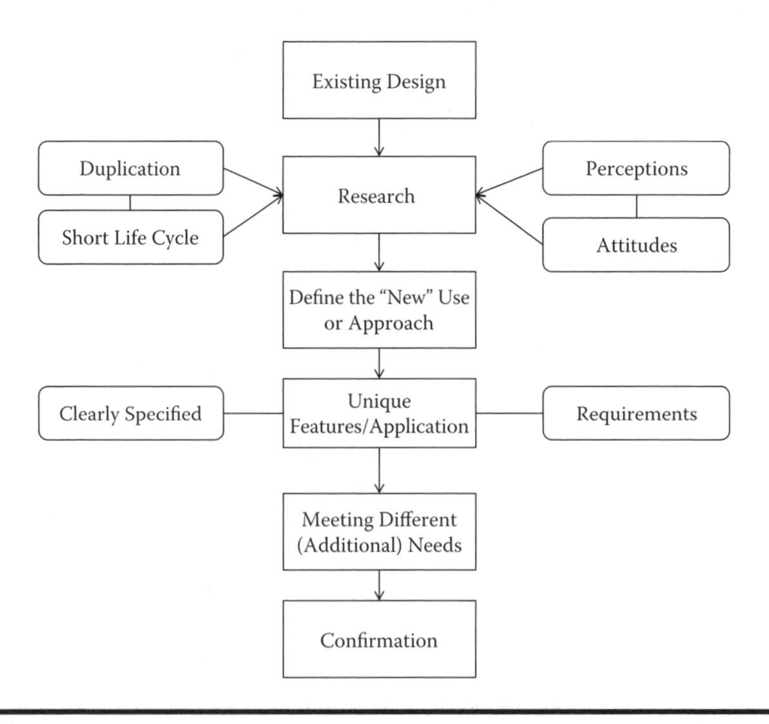

Figure 4.6 O—originality (new approaches, features).

group with an existing technology that has found a new application. To sell this feature will require precise, specific communications. The bulk of the research and money spent will be on the process of marketing or advertising the new approach. For those that recognize the benefit, new life is infused into the product, service, or technology.

The difficulty arises in convincing consumers/users that the new approach is innovative. Companies can easily make slight adjustments to the components, packaging, or service settings. Consider the ordinary mixer used for baking. By adding a new attachment the machine, one can now knead bread. The manufacturer need not change its production schedule, resources, or equipment; it need only to add a new attachment. Of course, there is a time required to develop, design, and test such an attachment, but consider the extended life cycle and the ability to advertise such a new product. Now, add a chopping device, identifying a new use for this very traditional machine. This example demonstrates the many new types of innovations that are possible. The difficulty that exists is to find resources that change or modify the item enough to be classified as a new use or new approach. Marketing and advertising personnel spend a great deal of time finding the differentiation needed to market this new item.

Summary

New features and approaches are a common method of "breathing life" into a mature product, service, or technology. It is a way of maintaining innovative excellence without the burden of creating new items on a frequent basis. As long as the business or organization is exceeding expectations of existing needs or identifying new (unfulfilled) needs, the item will remain innovative. As mentioned previously, new versions of iPads are innovative without being totally new. Competition is aiming to provide the same for their customers, but Apple maintains the upper hand at this point (it can offer new uses or a new approach and be seen as innovative). This is the simplest form of new and the one that is often misunderstood as innovative.

Discussion Questions

1. How original does a new design need to be to be considered innovative?
2. Identify each type of originality; list the pros and cons of each situation.
   a. What is the value of an item that is very original?
   b. How often can customers/users be convinced that a new application is innovative?
   c. Is it worth the time and effort to consider a new approach, given the ability of competitors to offer a viable alternative?
3. What are the strengths and weaknesses of each of the three new types of innovation?
   a. Identify a common set of criteria used to judge success.
   b. Which criterion carries the most weight or is most critical for judging success?

Creativity

All aspects of originality require a certain amount of creativity. Creativity is critical to initiate, maintain, and expand originality. At this stage, where items can be uniquely new—a new application with existing technology, or a new approach—creativity is critical. Creativity at this stage enables the individual or team to

- Visualize the product
- Identify critical characteristics
- Define the requirements
- Link technology with the needs of the customer/user

- Provide a mechanism to visualize the process, creating a source of information to maintain and potentially enhance the innovation

Remember what a sage (Dr. Theodore Levitt) says about creativity and innovation:

> The trouble with much of the advice business gets today about the need to be more vigorously creative is that its advocates often fail to distinguish between creativity and innovation. Creativity is thinking up new things. Innovation is about doing new things. (Peters and Waterman, 1982, p. 206)

At this stage, identify areas to address the need. Many creative techniques exist. When choosing one, consider the following:

- Is the technique a best match with the originality type?
- Will the team find value in the exercise?
- How difficult is it to control the activity in terms of the desired result?

Consider the type of technique used for idea generation. Be aware of each technique's use and application. Techniques such as brainstorming, affinity diagrams, or your own particular design may provide that added information to enhance the creative experience of each step.

Summary

At each step of the ENOVALE strategies model, creativity is needed. Creativity occurs in the mind; innovation acts upon the thinking and creativity process put into action. Creativity is a tool for innovation. Relying only on creativity (even in a structured fashion) alone is insufficient for sustained innovation success.

Discussion Questions

Begin with a particular identified need for your business or organization. Choose one of three options for innovation.
- What and when will creativity be required?
    - Name the creative techniques the team will use.
- Will the item have a short or long development/implementation time?
    - How does this influence overall acceptance from both the business and the user perspective?
- Are the customer/user needs addressed well enough for this innovation to be self-sustaining?

What is the overall chance for success for this item?

V—Verify, Value Building

Verification provides the final validation for the design to move forward. For the new use and new approach items, this should be a simple step to verify value. For the original design, implementation (proof of concept) must first take place before test and validation can follow (Figure 4.7).

Verifying the proposed concept leads to the next steps of verifying value (both internal and external). Establishing value requires input from customers, the organization, and the availability of resources. For all new designs, no matter the degree of originality required, there is a need to validate alternatives throughout the entire design, development, and completion stages. The alternatives provide opportunities and a fallback position. This type of thinking challenges the norm that states that there is "one best method" for implementing innovation. As innovation is a transformative process, this transformation will not always deliver predictable results. It may deliver unexpected results that require alternative actions. This is why a continual examination of alternatives is critical for overall success.

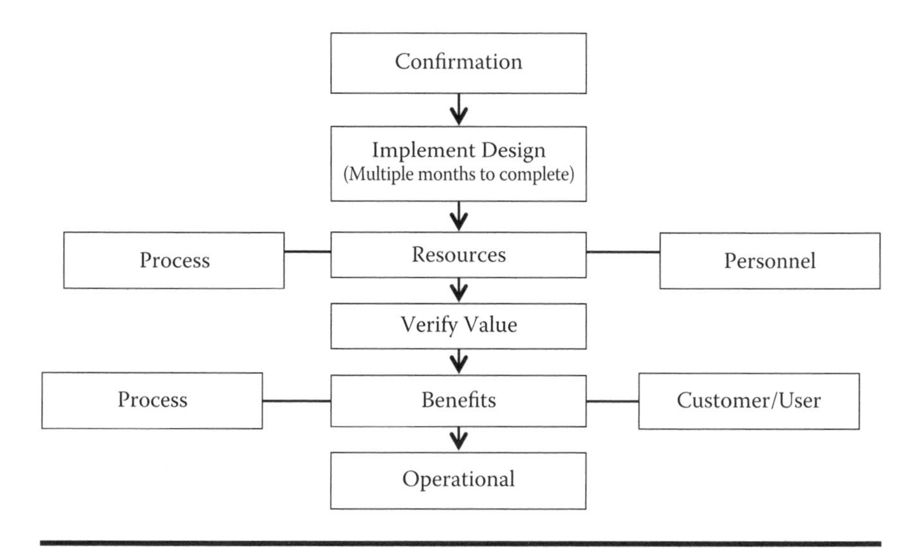

Figure 4.7 V—verify, value building.

Alternatives require an examination of both benefits and risks. Because these never occur equally, the process of comparing and contrasting benefits and risks provides a unique method of evaluating performance and user acceptance. Begin with the original and then alternatives; use a simple means to evaluate the following: benefits, risks, performance, and user acceptance (Table 4.2).

- Rate and rank criteria.
  - List the explicit benefit.

Table 4.2 Alternative Evaluation Matrix

Alternative	*Benefit*	*Risk*	*Performance*	*User Acceptance*	*Consensus*

    - Risk, performance, and user acceptance—use symbols to identify increasing, ↑, decreasing, ↓, and no change, ↔.
- Compare and contrast alternatives using the criteria.
- Reach consensus.

Example: Suppose a hospital administration wants to introduce a new vital signs monitoring device (Table 4.3). Evaluate alternatives from both the evidence available and the emotional value of the alternative—separate fact from fiction.

Summary

The amount of data available may be limited for evaluation. Therefore, it is helpful to evaluate designs before they proceed to full implementation. A simple comparison, such as demonstrated with the alternative evaluation matrix (Table 4.1), provides a simple overview of the original and possible alternative designs. Periodic evaluation of alternatives provides for not only a method to confirm the original design, but also fallback positions if one or more critical criteria change.

Discussion Questions

1. How worthwhile is it to evaluate alternatives?
2. What additional criteria would you add to the list of those used to evaluate success?
3. Consider the usefulness of alternatives. Validate fitness for use.
4. Consider the practical aspects of alternatives—what is their overall value?
5. What is the cost-benefit-risk ratio?
6. Can the alternative provide a secondary position if critical items such as resources change?

Table 4.3 Alternative Evaluation Matrix: Example

Original/Alternative	*Benefit*	*Risk*	*Performance*	*User Acceptance*	*Consensus*
High-tech device	Latest technology	Short-term learning curve, ↑	Excellent performance once implemented, ↑	Less noise, less invasive, ↑	Positives, with learning curve risks
Alternative 1: Simpler, lower-tech device	Existing technology	No risk, ↓	Nominal, ↔	No expected changes, ↔	No change
Alternative 2: Different manufacturer	Cost-effective, operates like alternative 1	Small risk given new equipment, ↓	Nominal, ↔	Noisy, can be evasive, ↓	Cheaper, performance good, noisy

A—Align and Adapt

Alignment and adaptation are consistent themes for ENOVALE and not detailed in this chapter. Unlike daily operations, innovations are not routine; they require a unique management process and the need to evaluate and validate at various stages. The management process consists of ongoing aligning of perceptions, tasks, communications, and emerging decisions. Upon reaching this stage, alternatives are either accepted or rejected. Therefore, the team needs to align itself to the final decision. This is the prime reason for sharing ownership. The organization owns all rights to the innovation that the team has facilitated. Most often, the team has little input into the final decisions. Unfortunately, this prevents their input and experience in developing and executing the decision to implement. It is noteworthy that decision criteria used by the team may not be identical to those used by executives in making the final decision. When venturing into new innovations, there is often a requirement for available funding and abundant resources. This goes along with the need for experienced individuals and having a process capable of turning a concept to reality (Figure 4.8).

Part of the process of alignment is to ensure the project assessments are subject to a reality check. These assessments can include, but are not restricted to

- Available cash (investment opportunity)
- Technology restrictions
- Available resources
- Available expertise and experience
- Existing process capability

Keeping a firm grip on reality permits the best results and those that seem to last the longest. Remember, reality for an innovation project is subjective at this stage, given the originality proposed. Reality may be a set of assumptions, not tried or

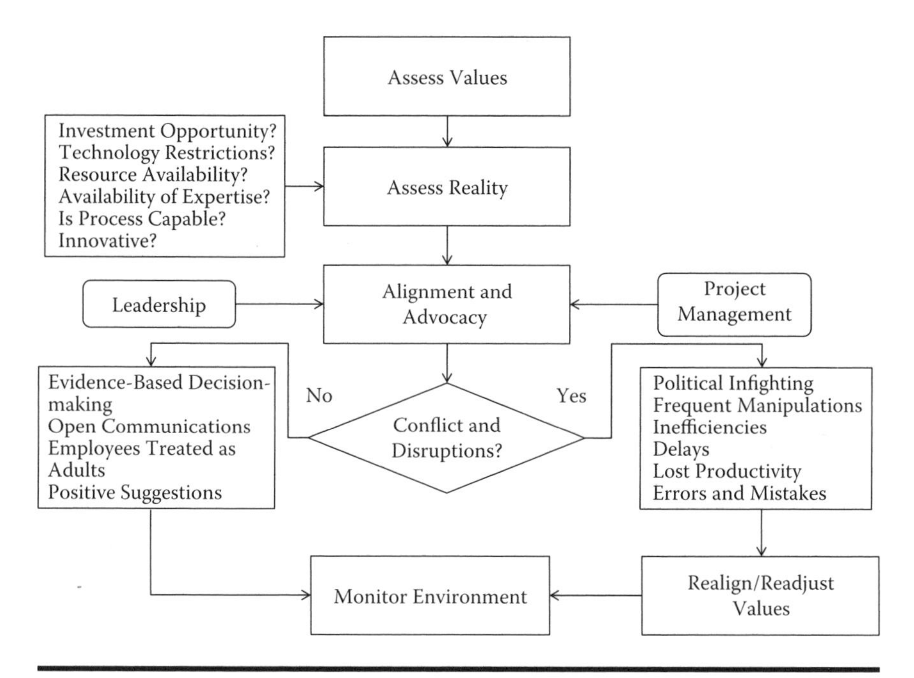

Figure 4.8 A—align and adapt.

proven. This is the reason for continuous alignment; as realities change, the team must adjust to the new situation.

Given that the approvals for the project are nearly complete and communicated to all concerned/required individuals, what is the process for individuals to align themselves to this final decision? It may seem rather unusual to ask this question; however, at the decision stage the number of changes, modifications, or suggestions has equally slowed. Especially with innovations that are new, constant modifications are frequent. When the modifications become interferences, problems will exist. Most often, the problem is one of leadership and project management. By this stage, it is important to solidify a leadership role, and an advocate role among top leadership. Leadership is important for managing the project. Individuals on the project team have just as important roles since they focus on executing directives and activities for the good of the project and its eventual success. The advocacy role is also critical. Top leadership must provide this advocacy role so that the project does not lose its importance or significance. Often leadership is excited about a project the first few months before it becomes a reality, but day-to-day activities limit their participation and interest. The importance of advocating for the project throughout its lifetime is one of direction and focus. If leadership misses this advocacy role, projects become entangled in political infighting, frequent manipulations, inefficiency, delays, and lost productivity. This all goes back to the old adage that "what gets noticed gets done."

It is important for these roles to be firmly established and fully functional. Each individual should be involved in the project as a leader, follower (team member), or advocate. To lead as an advocate for the project will require a focus on the benefits or advantages of the project and the decisions made to highlight these benefits. By communicating these benefits and advantages, the advocate can deflect any negative aspects that can occur that result in inefficiency and lost productivity. The followers will place their trust in their leadership when they know that the information is credible and the decision makers are dedicated to the ultimate goal, which is project success. It is important for leadership, in their advocacy role, to empower and support any changes or modifications that the team may find or recommend. Leadership must understand that the team is now in full control. The project team has the greatest exposure, experience, and interest in the project. As an advocate, leadership can provide constructive feedback, and limit the damage of any major modifications. Employees want leaders who will tell them the truth no matter whether it is positive, neutral, or negative.

How does management handle a negative decision? Inevitably, this will happen and employees will express their dissatisfaction. Ultimately, there will be times that employees may not agree with the leaders' decisions. The question then is, how do these leaders align employees and team members to this unpopular decision? The following are some suggestions for alleviating this problem:

- When employees oppose a decision, leaders provide evidence to the contrary.
- Communicate the advantages of the decision; recognize the information provided by those who oppose the decision. It is better to tell someone that you do not agree with them and present a factual case than to use pure opinion and feeling.
- Trust employees as adults. Truthfully inform them of the decision. Trust them to conform to the decision, but always permit some time for acceptance.
- Provide a mechanism for positive suggestions or solutions.

Employees can easily accept even an unpopular decision when they know their voices are heard and the evidence demonstrates a different (and better) result than the employees thought or reasoned as possible.

Summary

In summary, this section aligns individuals, team leadership, and executives to a final decision. The purpose is simple: get everyone to adapt his or her behaviors and attitudes to the project decision. Unpopular decisions may require extra effort and time for people to adjust. Adjustment can be short or long, depending upon the intricacies of the project and its result. Alignment at this stage focuses on moving forward the project from outcome to concept to reality. Although many businesses and organizations are adept at project management, innovation management requires a unique set of steps to succeed. For innovation projects, the checks and balances developed during the ENOVALE solutions phase determine overall viability. If a team has followed this strategy, then the project should develop through its planning phase and be ready for implementation.

Project success hinges on leadership and advocacy. Leadership provides the resources, supports the project, and directs the high-level efforts. Advocates not only support the project, but also address conflicts and disruptions. Both groups provide a clear message dedicated to project success, while alleviating barriers.

Discussion Questions

1. Describe the steps you would use to align a poorly functioning team?
2. What can management do to alleviate the problem of misalignment?
3. How does an advocate work? Does it have a place in the overall project?
4. What strategies can be put in place to alleviate or minimize misalignment?

L—Link to Performance

This phase is very similar to the one that precedes it. Good planning skills in the beginning should easily accommodate this phase; this phase tests whether the process will work as expected. This is certainly time to bring in your organization or company's business, marketing, and financial individuals. Obviously, the new form of innovation must conform to corporate expectations, requirements, and objectives. Be very careful if using this stage to review alternatives—many opportunities are lost based solely on profit and loss (Figure 4.9).

When originality is critical, this stage will be more important and more dynamic as well. Something unique and distinctive will have financial measures associated with it (after exiting the ENOVALE solutions phase). These financial measures will likely be profit and loss related. Although profit and loss are quite important, the notion of measuring many characteristics, whether they be human related or process related, or results and customer related, is part of the innovation

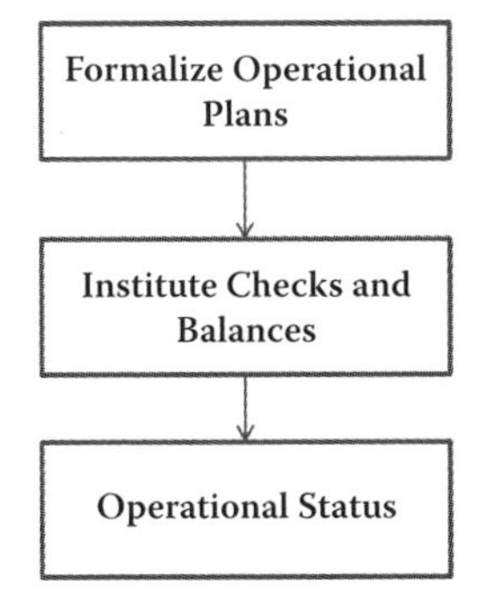

Figure 4.9 L—link to performance.

process. Again, for new designs, many of the measurements will most likely be limited to experience or opinion. At this stage, there is a true need to link to existing operational performance measures within the organization. It is important to remember that organizations rely heavily on criteria that are not innovation based. Accounting systems support month-to-month activities, rather than innovation project metrics. Hand-in-hand with these measures are those that judge performance, which is a key factor in innovation—not to mention the many intangible characteristics/measures that innovation requires for a thorough assessment.

Once the first project is developed through ENOVALE, there will be a growing list of opportunities for which metrics are readily available. New projects require additional information and many performance metrics will require development, testing, and validation. Even if little information is available on financial performance, or even estimates of success, experience has shown that a well-prepared team following the steps discussed above will be able to make a compelling case to management.

Summary

In general, new innovations require a set of metrics that may need development to assess project success. For truly original products, services, or technology, metrics may not exist, and only experience and judgment suffice to evaluate performance. Project teams may find that accounting systems do not have the fine level of detail needed to identify complete project savings. This is where creativity and resourcefulness come together to provide a measure that permits evaluation of success. Over time, the business or organization will accumulate these for use in any project decision.

Discussion Questions

1. What expectations exist for performance for this new type of innovation?
   a. Is there an existing baseline?
   b. If not, can one be created?
      i. How would you establish the baseline?
   c. How would you expect the customer/user to respond to this performance?
2. How is competitive advantage (value) measured and evaluated?
   a. What are the distinctive characteristics that give the item its competitive advantage or value?
   b. What is unique about the product that ensures a long(er) life cycle?
      i. For items that are a new approach, what is the life cycle?

E—Execute the Decision Plan

This last step is quite unusual, since the evaluation concerns only the decision, rather than the design or concept. Often decisions use a collection of evidence, emotions, and feelings. Available evidence (data) often results in the best decision. Rely on facts and evidence blended with experience and sound judgment. For this type of innovation, a strong leadership role

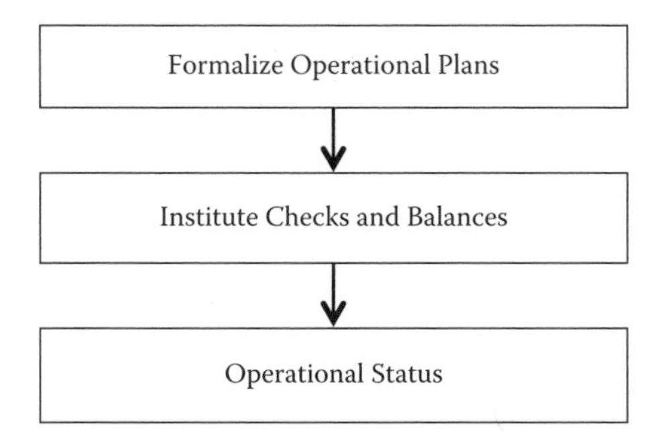

Figure 4.10 E—execute the decision plan.

is needed, and for this reason we call members of the leadership team an advocacy group. Remember that over 90% of all decisions made in businesses and organizations rely on feelings and emotions. The problem is that only 10% use evidence-based results. Managers need to use evidence-based decision making when it comes to innovation (Figure 4.10).

Consider the following criteria for evaluating a final decision:

- Is the decision well accepted?
- Is the announcement welcomed or contentious?
- What was the attitude of the team at the time the decision was made (supportive, negative, noncommittal)?
- Does the decision agree with the consensus of those who evaluated the product, technology, or service?
- Were facts and evidence useful for more than 50% of the criteria in these decisions?

Negatives with any of the questions posed will surely result in a potential failure mode. One key to effective decision making is having evidentiary information supporting experiential decisions. Information can only come from a functioning measurement system. Therefore, to evaluate the final decision, make effective use of existing information.

Finally, establish the operational plan for the innovation. At this stage, the outcome is viable, the concept and design have been proven, and final preparations are proceeding on transforming the item (now generally a prototype) to a more usable format for rollout. Develop a project plan for this innovation, including tasks, responsibilities, and sufficient measurement and evaluation for all decision points. Assess risks, consequences, and alternatives. Operational guidelines evolved into the operational status that provides the plan for success.

The benefits of good planning include some principles that dictate the need for alternative practices, thinking, and response. The ENOVALE philosophy underscores the importance of the human being driving innovation. Humans do err (make mistakes), and therefore risk and consequences are always part of the total picture. Several good, viable innovations were never implemented simply because they lacked effective planning.

Summary

The final step is just a microcosm of all seven steps that emphasize planning, measurement, a fair evaluation, and negotiation. Creating something original is never simple, but the rewards are great. Innovation is not easy, but once you have moved through the process, the next time will be shorter and less treacherous. The purpose was to demonstrate how to achieve operational status through assessment, evaluation, and measurement.

Discussion Questions

Take a simple decision and apply the following criteria to that decision:

- Examine the procedures and communications used to present the decision.
- Examine the results and consequences.
- Determine the level of satisfaction with the decision.
- Discuss warning signals present for negative decisions.

Identify three additional criteria needed to evaluate a decision.

Chapter 5

Improvement Projects

Introduction

Taking an existing product, service, or technology and making these items "better" far outweighs the need for new items. Improvement is the simplest innovation to recognize, and the one that most influences competitive advantage. This need to improve should be a mantra for all businesses and organizations.

Fredrick Taylor, at the beginning of the twentieth century, recognized the need to improve in a systematic manner. He was the first person to explain why businesses needed to be more efficient and effective, and his "scientific management" method still exists in numerous forms today. Taylor was concerned that, over time, machines and humans become less efficient and performance decreases. Management could provide a best method for operations, and then train the workforce to follow these explicit directions. The intent was to keep production steady and processes efficient. In the 1950s, the Japanese applied these best practices in an attempt to rebuild their economy after World War II. The Japanese applied these methods, as taught by Doctors Juran,

Deming, and others developing the field of quality assurance. While Japan applied these best practices, the United States returned to its practices applied before the war. Over time, these practices led to inefficiencies, poor quality, and ineffectiveness. During the 1980s the same Americans that so diligently taught the Japanese again began teaching managers in the United States these “secrets to success.” The field of total quality management arose and was popular for approximately 10 years. When the luster left this technique, Six Sigma and Lean techniques became popular in the mid-1990s, and still are today. Success, using these methods, has had mixed results due to the fact that management generally wanted to measure progress only in terms of costs and benefits. Many costing (accounting) systems are inadequate to identify and “bank” these project savings, and therefore many organizations never achieved the success with the method that was possible.

The need to sustain and increase performance is driving businesses and organizations today. Combined with an emphasis on innovation, businesses must exceed or improve their performance to remain competitive. Innovation generated by improvements results in a product, technology, or process that is significant, something substantially better than what previously existed. The difficulty is that this concept lacks definition and specific meaning. Chapter 2 presented an innovation primer for those who need to understand its meaning and how individuals judge it. For this section, the innovation theme is improvement, that is, making something better than presently exists. There are two ENOVALE® applications: one is for improvement of some item that needs to outperform (is presently performing at or near existing expectations), and the second is for items underperforming expectations. The keyword is *performance*, and how it must be better than what presently exists. Figure 5.1 outlines the similarities and differences. The shaded areas are where strategies differ.

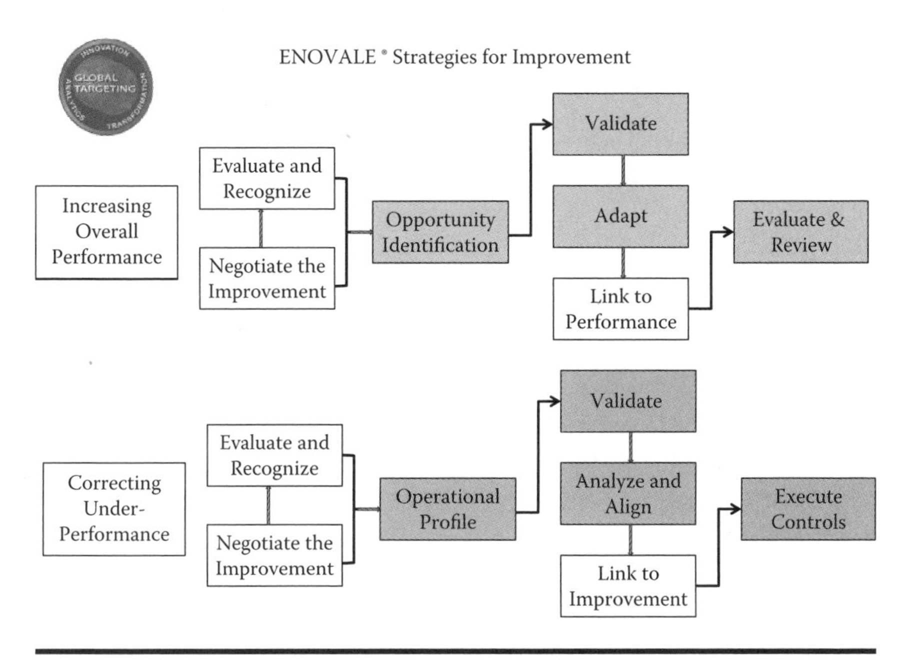

Figure 5.1 Two-tiered improvement project strategies.

Performance below Expectations (Underperformance)

One of the critical keys to successful innovation is outcome (product, process, service, or technology) performance. Performance is any measure of achievement. Performance could be as simple as doing a task faster than expected, or as detailed as performing a series of complex algorithms successfully. Performance is a measure of success. Expectations go hand in hand with performance. Satisfaction increases (as does purchase behavior) when expectations are exceeded. Therefore, performance is a key to recognizing innovation. If an outcome (item) does not perform to customer or user expectations of performance or meet their specific needs, it is not innovative (much less acceptable). This assumes, of course, that the customer or user has previous experience and knowledge with the item. New items, for instance, have no history or performance from which to evaluate. For improvement, however, one has to have knowledge and experience to judge the item as better than its predecessor does. When performance is erratic, it will not meet expectations or satisfy existing needs. The purpose for innovation is to improve the product, process, service, or technology to meet performance expectations and existing needs. Therefore, the team desiring an improvement must first decide how and why the item is underperforming.

The ENOVALE process, for an outcome that is underperforming, stresses first the ability to understand the reasons for underperformance, and then remedying the situation. What has caused a change in purchase behavior is that performance fails to satisfy due to erratic causes or unexpected consequences.

Performance above Expectations (Overperformance)

More than likely most innovation projects will attempt to improve performance even if performance does not meet expectations. Most providers are looking for opportunities to increase the performance of their product, service, or technology. Improving the performance is certainly much less expensive and time-consuming than inventing a new product, service, or technology. Rather, like in the previous section that focused on improving performance due to causes or reasons, this approach assumes that performance is as expected, but increasing that performance will open new opportunities for growth, sales, and competitive advantage. It is natural that most providers assume their item is performing at or near the expectations. Therefore, this approach would most likely be the first choice of many providers who assume that increasing performance will provide additional benefit to the company or organization. The danger in this assumption is that if the item is performing inconsistently or erratically, improved performance may never occur. The business must focus on consistent performance. Improvement also assumes a second assumption: that customers or users have their needs satisfied. If customers (users) need increased performance, then they will search for a more appropriate substitute. Identifying the best approach is the ultimate success during the outcome development, validation, and verification stages.

E—Evaluate and Recognize

The first step is to evaluate performance by establishing a baseline. Creating this baseline dictates which ENOVALE path the project will follow (Figure 5.2). Creating a baseline requires four unique elements.

1. Identification of a measure that best measures performance
2. Identification of a proper measuring instrument or device

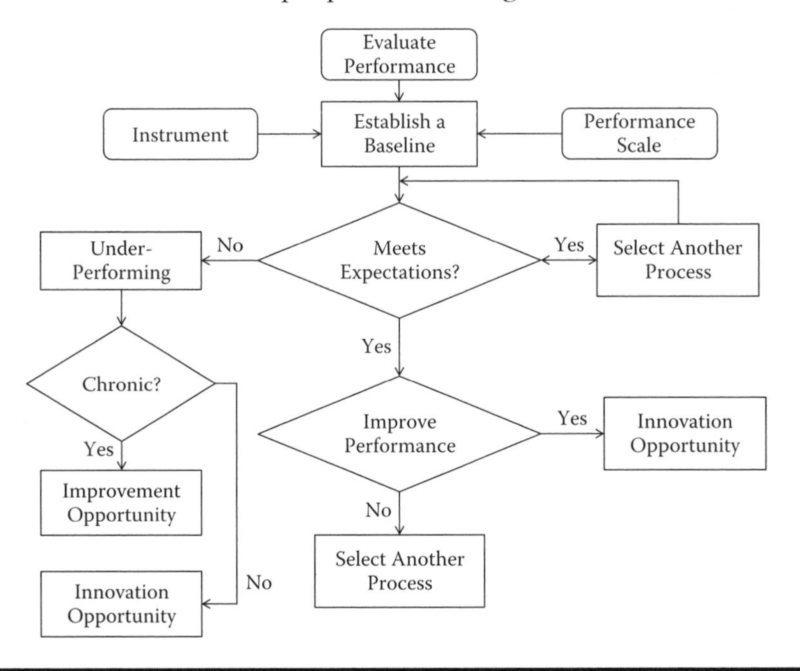

Figure 5.2 E—evaluate and recognize.

3. Creating an appropriate scale to rate the performance
4. Consistent, accurate, and timely observations

The baseline is the expected performance of the product, technology, or service. Expected performance may involve both tangible and intangible measures. Monitoring and control require precise measurements. Intangible measures, such as applying sensory information, are critical to judging performance. Intangible measures require a standard from which to judge performance and rater (evaluator) consistency. Measurement capability, reliability, and maintainability are excellent resources to manage the measurement system.

One such device that encompasses both precise measurement and human sensory elements is a simple flow meter used to control the amount of intravenous solutions entering a patient. For many years, a simple (two-piece plastic) roller-type flow control metered the rate of intravenous solution infusion (many people will remember the "drip" of fluids). Now, infusion pumps have replaced the flow control devices. This simple meter was precise (a true innovation) for its day. Nurses were especially fond of the simple device, as it was easy to control the flow rate. However, compared with today's technology, it is imprecise and less accurate than electronic devices. The point is that the feel and ease of use (a human response) were as critical as the actual flow meter control properties.

Although many people strive to collect data associated with performance, often the measuring or measurement device is a key to determining if the item is performing to expectations. In collecting the data, ensure they are simple, straightforward, and readily available. The simplest measurement is often the best. Reliance on only mechanical devices removes the dimension of human sensory information. For example, long before the "engine" light appears on a car, the senses will already provide a signal that something is wrong. Sensory information is powerful but requires a standard and a method of testing rater (evaluator) repeatability.

Products require measurement that may involve both mechanical and human evaluation. Often, part of the improvement process is the acquisition of new technology as a means of improving adjustment and control. If performance is poorly measured, then improvement may become an unrealizable goal.

Services require both standard and sensory or even a psychological response (feelings, attitudes, or perceptions), which all require a baseline. In all cases there needs to be a method to evaluate performance. How an individual judges innovation directly relates to performance. If you cannot measure performance, how can it ever improve?

Oftentimes the difficulty with measurement is with the scales used to evaluate the measurement. Scales can be complicated or they can be simple. There are times when a simple yes or no is appropriate to evaluate performance. However, most times there is a need for a more descriptive measure, which captures more information about the outcome or item itself. This type of data, generally called continuous data, is defined along a continuum ($-\infty$ to $+\infty$), measured to an infinite degree of detail. Probably the most consistently used continuous set of data is that associated with time. Time is measured in degrees of sensitivity from light years to nanoseconds. That is, there is an infinite amount of discrimination one can apply to defining a period of time. Scales can also be descriptive. In services, the scale that describes how well you agree or disagree with the statement describing performance is often perceptual. Surveys provide one mechanism for measuring performance. Whatever you use to evaluate performance requires a scale that can easily differentiate between what is and is not acceptable. When a product, process, or service performs below expectations, there is a need to improve. Figure 5.3 describes the general measures of scale and gives some simple examples. Scales are critical for understanding and evaluating performance. Readers need to establish measures and scales to evaluate performance. Begin with the

Categories of Scale	Description	Examples
Nominal: Unrelated categories that represent membership or nonmembership.	• Attribute or sensory data • Grouping/sorting • Yes/no, pass/fail	• Categories • Labels • Classifications
Ordinal: Ordered categories with no information about distance between categories.	• Count data • Ranking • Surveys	• 1st, 2nd, 3rd • Strongly agree to strongly disagree • Alphabetic order • Rank order
Interval: Ordered categories with equal distance between categories, but no absolute zero point.	• Continuous data • Most common scale • Use arithmetic with caution	• Temperature scales • Dial indicator
Ratio: Ordered categories with equal distance between categories with an absolute zero point.	• Continuous data • Proportional relationship • Most forms of arithmetic apply	• Air pressure gauge • Velocity = distance/time • Ruler

Figure 5.3 Categories of scale.

scales and measures that are simple to extract and easy to understand. The purpose for these measures and scales is to collect the best data to establish the baseline performance for the outcome (or item) chosen for improvement. Many business organizations have a detailed set of data that measure the performance of various products, processes, or services. However, in addition to the data that are collected and evaluated, you can expect that a minimum of 20–50% (estimate based on years of experience) of additional data are required to truly understand and judge performance. For those critical measures, not presently collected or evaluated, we recommend you begin with a nominal scale to evaluate performance. The premise is to keep the measurement simple and the scale specific to performance that requires evaluation.

Improvement requires a measure of performance that is not based solely upon opinion or feeling. If someone says, "I think this is not performing well," this is an opinion. We need information (data), frequently called evidence, to decide what and how to improve an item. When measuring an item, begin with the baseline and a scale of measurement that fits the item's particular measure of performance. If the item meets your criteria for performance, then move on; if not, choose another. Whatever scale of measurement you decide upon, be sure that you collect at least some numerical data on the item as they relate to performance.

If data are available, then proceed to chart or graph the data. A visual inspection is critical in providing a method to evaluate the data. Plotting a set of data may seem to be unnecessary when discussing performance. However, the visual display of your data provides a wealth of information not only about the item, but also about the recurring patterns and trends in the data. You can easily plot or chart using programs such as Excel. Once the data are plotted or charted, you can use sophisticated probabilistic guidelines (if you want) to determine if the data are stable or erratic. This assumes that the data you have collected are graphed or plotted in time order. If so, calculate an average or median and place it on the chart or graph. Look at the data as they vary from the median or average. One such helpful tool is the run chart (see Figure 5.4 and Appendix A).

Run Chart

Run charts are a simple but effective tool for diagnosing inconsistencies in a set of continuous data. Many businesses (especially in the process and manufacturing areas) use sophisticated statistical tools such as control charts to accomplish the same objective. The purpose is to measure performance, detect changes, and deploy corrective action to maintain consistency. Run charts are a simpler method to detect large changes in a measured process. If the opportunity

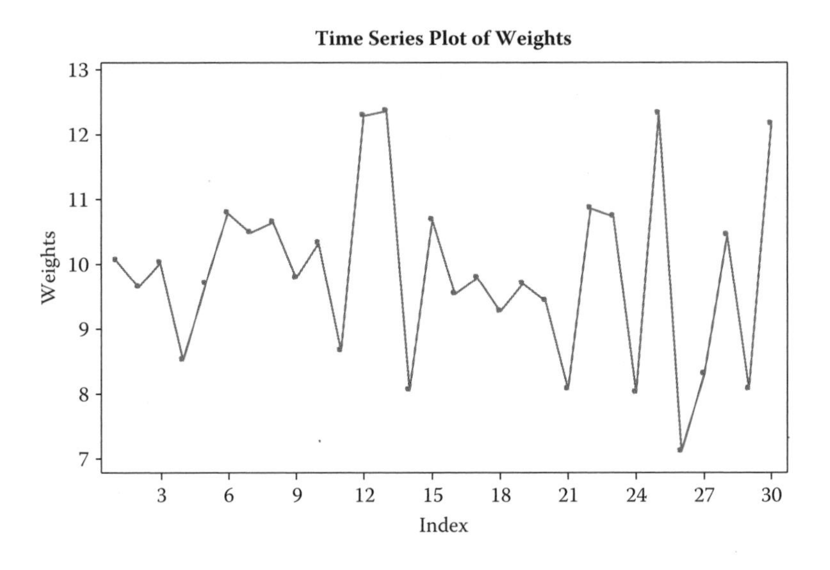

Figure 5.4 Minitab 14 output—30 weight measurements.

exists to measure such data, even with small sample sizes, the run chart is an excellent tool.

When using the run chart (with a median—solid line), look for consistency in how the data vary from the median (average) and look for randomness (no specific pattern or trend) of the data, as this is a signal of consistency. Appendix A discusses the procedure to evaluate trends, cycles, or erratic swings in the data. A trend, cycle, or erratic swing is indicative of inconsistency. This indicates that the data cannot maintain their baseline and therefore will exhibit patterns of inconsistency. Just because the data vary above or below the average or median does not mean that they are erratic or inconsistent. The example data (Figure 5.4) do vary, but consistently around the median or average. Therefore, a predictive analysis is possible, as the data vary around the median. Any points that seem inconsistent (out of place) are specifically due to reasons that change the natural rhythm of the data, causing either a trend or exaggerated cycle or periods of large or very small inconsistencies. If you do not have the ability to look at a chart or graph statistically, then the visual inspection is extremely helpful. Look for anomalies or unusual points in the data set and find the reason or cause for these instances of erratic performance. More often than not, these will provide clues as to the cause or reason for the inconsistency in performance.

Creating the baseline enables the team the ability to assess expectations. Either the item meets expectations (and can improve) or it does not meet expectations. If the item cannot meet expectations, then innovation is not possible, but improvement is warranted. If performance is chronic (an underperforming at times), then there is an opportunity for an innovation project.

Identifying a measure of performance additionally requires the establishment of a single success criterion or a set of success criteria. A success criterion is simply a description of how the item will perform. For example, the wait time for a meal in an expensive restaurant is far greater than the wait time at a fast food restaurant. The difference between the two restaurants is the fact that the expensive restaurant adds dimensions of service (performance), including the service itself, the food, and the ambience. Therefore, improving the wait time in the expensive restaurant would involve more than one success criterion. Obviously, the simpler the baseline, the easier it is to improve performance. The baseline must consist of available data. Baseline data can be quite revealing, specifically if something unexpected or unusual occurs. Changes in the natural rhythm of the data are indicative of causes and reasons for problems that directly relate to underperformance. Rather than ignoring this type of data, it is best to collect this information to provide more a realistic baseline performance for the item under consideration.

Summary

Performance is the critical element of the evaluate and recognize phase. Without measuring performance, innovation is impossible. Measuring and evaluating performance is often the most difficult aspect and greatest barrier to innovation. Unless an adequate measurement system is in place, improvement is difficult and innovation nearly impossible. Use the measurement system to establish a baseline and evaluate performance. Collecting and evaluating data is critical to measuring performance and innovation.

Discussion Questions

Consider a possible improvement project; define the measurements needed to evaluate performance.

1. Are the data available (measurable) (if yes, go to question 3) or must they be created?
2. For the data that do not exist, create a measurement and measurement scale.
   a. Consider data type. Is there a measurement that lends itself to visual analysis?
   b. Can you construct a chart, graph, or table with these data?
   c. What scale best describes these data?
3. How will you collect these data?
4. What would be the expected baseline?
5. What will a data analysis accomplish?
   a. Will it measure and judge performance?
   b. Will it provide information on changes to the process?
   c. Can it track underperformance or overperformance?
   d. Does it provide a means of controlling the process?

N—Negotiate the Improvement

The possible opportunity requires the input of all those associated with the project. Before selecting a project implementation team, leaders, managers, and operational and technical

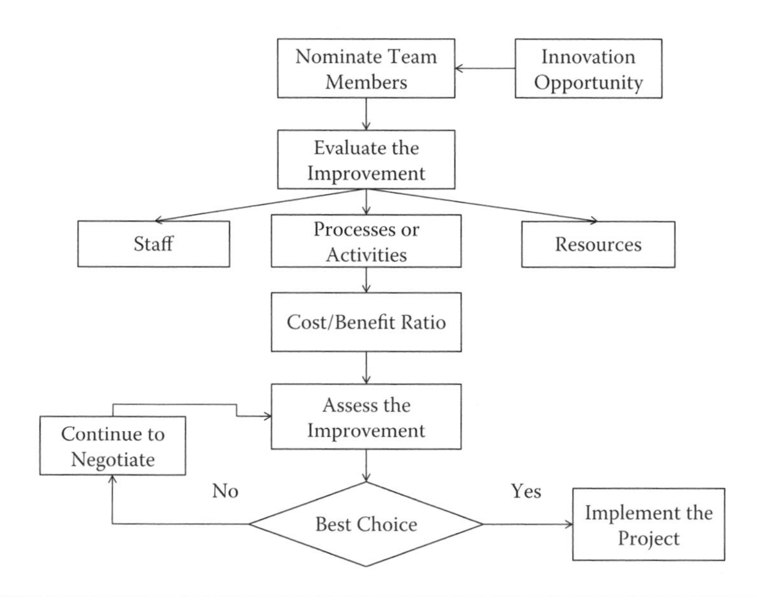

Figure 5.5 N—negotiate the improvement.

personnel should meet to decide the fate of the project. Sufficient information should exist to aid in this decision. Selecting the best candidates for the project team also occurs at this phase. Use the ENOVALE solution strategy to select the best candidates (Figure 5.5)—nominate team members.

Identifying the best candidates to work on the improvement relies heavily on those who easily recognize innovation as improvement. The importance of nominating the best individuals is highlighted by the fact that these individuals are aligned and adapted to work on an improvement project. *Chance or Choice: Unlocking Innovation Success* provides an excellent discussion on whom and how to select the best individuals. If additional questions exist concerning the best candidates for the project team, we suggest you consider an e-mail to us at contact@globaltargeting.com.

One critical aspect for this type of innovative improvement is the ability to negotiate within the project team and to manage based on information gathered from previous phases. Generally, people working on this improvement project will view it from different angles, and therefore propose different (and varying) solutions. Although one solution may be better than the next, we begin the process of examining the reasons and causes for the loss of performance before we implement a particular solution. Therefore, it may be necessary to negotiate how and what the improvement outcome will be for this project. This enables the business organization to use not only evidence-based management, but also the experience and wisdom of the organization in choosing the best improvement project. The goal is to create an innovation recognized by customers and uses alike. Recognition by the customer or user that the performance has improved is the key to this type or theme of innovation. We expect that the need driving innovation will not change, but that the performance of the item will improve.

The negotiation process should consider a number of elements before making a selection. First, evaluate the process in terms of how it influences (affects) staff, processes, or resources. The business must convince itself that the improvement is worthwhile. Avoid those instances where short-term fixes increase performance. By fixing a problem (selecting a solution and making it fit), the organization may or may not identify the cause or reason for the problem, and therefore only delay future inconsistencies, reduced performance, and loss of success. In addition, focus on long-term performance due to those recurring problems that cause inconsistency in performance. Use available data, experience, and good judgment to assess if the improvement will go forward. Rather than just looking at a single benefit such as cost reduction or benefit enhancement, consider long-term value as one characteristic used to evaluate performance. Outcomes that generate value over the long term result in consistent performance. Implement the project when achieving agreement on what will improve, how much it will improve, and whether the improvement is experienced by the customer or user.

As a reminder, our focus is on innovation, which is different from everyday activities. We would never expect a process that generates day-to-day results to be one that is capable of being innovative. Innovation requires a transformation; when we remove the obstacles that prevent performance from meeting expectations, we can then begin the improvement process. It is noteworthy that discussions involving replacement are not synergistic with improvement. Improvement takes an item and makes it better. It is as simple as the last statement.

Summary

In summary, this step finalizes the decision to move forward. By nominating the best candidates for the project, the organization ensures itself a fair and honest evaluation. Let the team negotiate the outcome and its milestones and objectives. The negotiations should center on the benefits that the innovation will bring to the customer or user. Negotiating the improvement from a business decision, and a customer purchase enhancement, is useful for selecting the most appropriate projects to develop. Negotiation provides an opportunity for alignment and the ability to close the gap between expectations and perceptions.

Note: At this stage of the ENOVALE process, there is divergence between the objective of improving performance and that of improving the process to meet performance expectations. Improving the process to meet performance expectations is described first.

Discussion Questions

1. As a team, how would you nominate a new team member?
2. What one criterion would the team use continuously to evaluate a potential innovation project?
3. What are the key resources needed to move the project forward?
4. At this stage in the project, what elements would you consider with regard to the cost-benefit-risk ratio?

5. How does the team work with management to make the best choices?
6. What issues may remain after making the decision to implement or not to implement a project?

Improving Underperformance to Meet Expectations

O—Operational Profile

The first portion of this step identifies the reasons why an item or component needs to be improved. An item needs improvement, from an innovation standpoint, because it is underperforming (Figure 5.6). Performance is less than expected. Consider the introduction of Windows 8 and its perception in the marketplace. Innovative software performs better than expected while continuing to meet a whole host

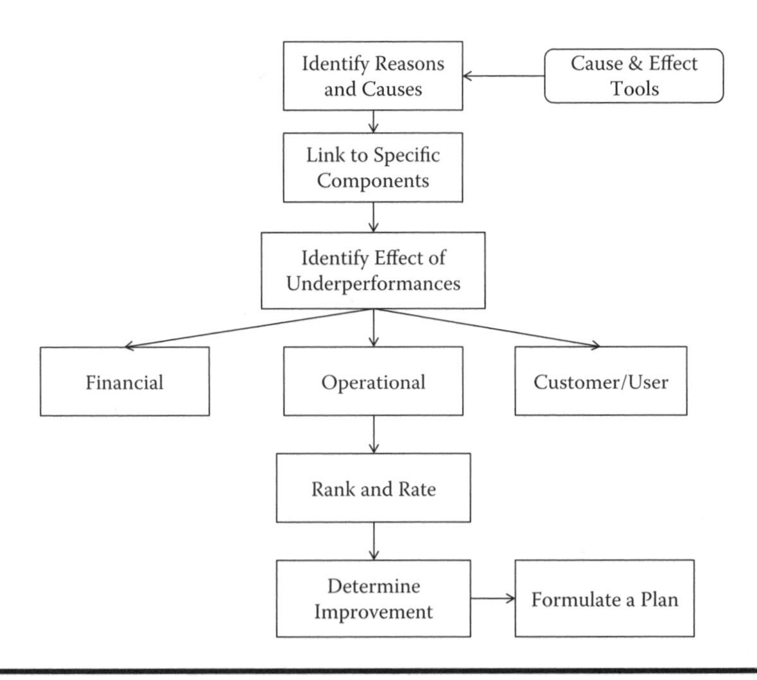

Figure 5.6 O—operational profile (underperformance).

of needs. If the software performs as expected, then it would need a specific differentiating factor in the marketplace. By now, the next generation of Microsoft is not a one-of-a-kind innovation, but merely an improvement from the previous version. Customers/users will judge performance whether or not it continues to use existing systems/software, or at least a variant of that system. Even if the software offers many new features and fixes, if it does not correct many older problems, it is not innovative. That is certainly a concern for Microsoft as it competes against products such as Apple, which has proven its innovativeness, given the fact that it not only performs well, but also meets new needs. For Microsoft to succeed, the software needs to excel (no pun intended) in performance against the expectations it has set for itself. In fact, this is a good example of how the consuming public may not identify internally driven innovations. This is why the process of identifying innovation is time-consuming but specific, so that product, service, or technology offerings are clearly recognized as innovative. If Microsoft understands the concept of innovation from an individual perspective, then we can expect new versions of Windows to not only meet expectations, but also excel in performance compared to its previous version. Customers and users can see through a veiled attempt to introduce something as improved when in fact it is not improved.

As stated previously, underperformance is due to a specific set of reasons or causes. It is the responsibility of the producer to eliminate these reasons and causes and return performance to expected levels. The assumption is that the company or business has conducted research to determine that improvement in the item will be beneficial to the organization. If this improvement experienced by customers/users is real and meeting existing needs, then we have true innovation. To begin the process of returning performance back to expected levels, there is a need to examine, rank, and consider the reasons for the underperformance. Consider the following:

1. What components/elements are underperforming?
2. Why is the underperformance so critical to the overall acceptance of the product, service, or technology?
3. Who and what does the underperformance affect? Be specific here.
4. How does underperformance affect the user going forward?
5. How much does the underperformance affect the bottom line?

Upon answering these questions, step back and reevaluate this situation.

When identifying underperformance, examine the reasons and causes of the phenomena. There are many tools available to determine causes and reasons for underperformance. The first step is to assemble the team, examine the data analysis that confirmed the underperformance, and determine possible reasons and causes for the loss of performance. One helpful tool to use is that of brainstorming. Brainstorming enables the team to create a set of causes and reasons for a specific problem. Brainstorming is a tool that is widely used for many improvement methodologies. The largest concern with using such a tool, such as brainstorming, is the tendency to think that the technique creates a chaotic response in regards to idea generation. Brainstorming is a structured process of collecting ideas and then refining these ideas until the team has reached consensus on a set of causes/reasons for a problem. Other useful techniques, such as an affinity diagram, may also be helpful for identifying reasons and causes.

The objective of the brainstorming or other idea generation session is to identify key reasons and causes why the outcome is underperforming. Begin with collecting ideas regarding what are the most common and prevalent reasons and causes for the underperformance. During the collection process, do not permit comments or opinions to override causes or reasons. At this stage, it is important to record as many causes as possible.

The next step is to identify a key outcome that is descriptive of performance. The outcome becomes an effect, given that the causes affect performance. A cause-and-effect diagram is then constructed. The cause-and-effect diagram (Ishikawa diagram, the originator of the technique) visualizes the relationship between the cause of underperformance and the effects of underperformance. This tool, often called a fishbone (or cause-and-effect) diagram, is useful for examining causes without applying a ranking or rating scheme. Identify the main categories of causes or reasons and list these as identifiers on the cause-and-effect diagram. Standard categories are people, materials, machines, procedure (methods), measurement, and the environment. Arrange the causes that best fit under each specific category. From here, the team can then rank and determine which causes are most in need of improvement.

A simple visual tool, the cause-and-effect diagram (Figure 5.7) provides a mechanism of classifying causes. In the

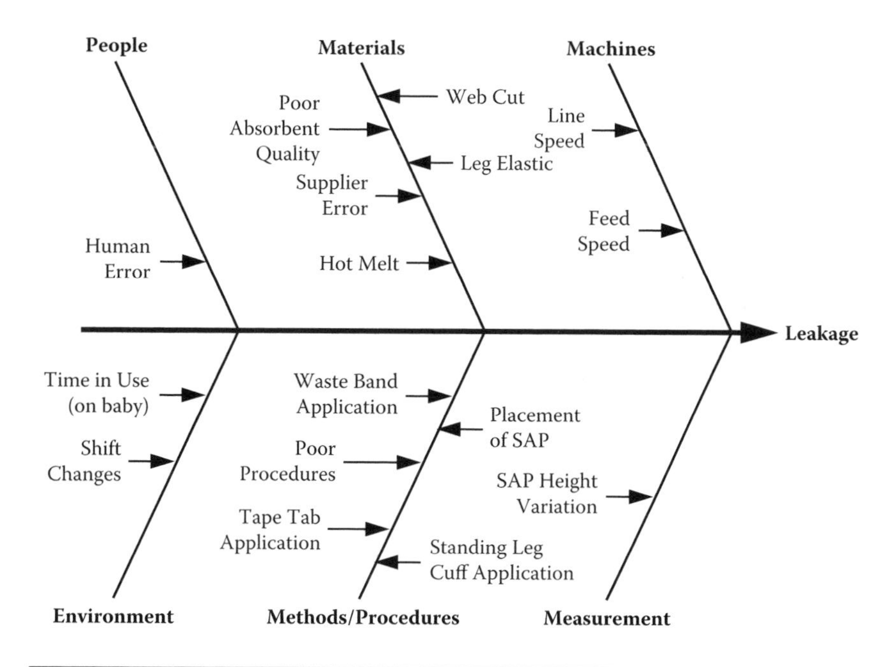

Figure 5.7 Example cause-and-effect diagram.

example, the effect is leakage; the causes are problems with children's diapers.

The cause-and-effect diagram is a visual diagram; it does not assign ranking or rating to the causes listed (Appendix C). One additional tool, the cause-and-effect matrix, is available for more complex processes, where requirements may temper the effects. The cause-and-effect matrix is a Microsoft-based Excel tool that can calculate a score to measure the degree of association between causal elements and more than one effect.

Figure 5.8 demonstrates a tool to determine which cause or reason accounts for the greatest loss of performance. The next rational step is to find a technique that evaluates the cause-effect combinations. The cause-and-effect matrix (Figure 5.8) does assign rank order to the cause-effect

Cause and Effect Matrix																		
Rating of Importance to Customer (needs) or Project Objective (requirements)																		
			1	2	3	4	5	6	7	8	9	10	11	12	13	14	15	
Choose either a Cause or Process Step			Need/Requirement	Need/Requirement	Need/Requirement	Need/Requirement	Need/Requirement	Need/Requirement	Need/Requirement	Need/Requirement	Need/Requirement	Need/Requirement	Need/Requirement	Need/Requirement	Need/Requirement	Need/Requirement	Need/Requirement	Total
	Process Step	Cause																
1																		0
2																		0
3																		0
4																		0
5																		0
6																		0
7																		0
8																		0
9																		0
10																		0
11																		0
12																		0
13																		0
14																		0
15																		0
16																		0
17																		0
18																		0
19																		0
20																		0
																		0
																		0
Total			0	0	0	0	0	0	0	0	0	0	0	0	0	0	0	

Figure 5.8 Cause-and-effect matrix template.

combinations. Appendix F contains instructions and interpretation of this chart.

Once the team has rated the causes or reasons for the loss of performance, the next step will be to collect information on the critical causes to determine their overall effect on performance. The most important element of this stage is to observe and confirm that particular reasons and causes that affect performance in such a way do not change customer or user expectations. Focus on capturing information related to the causes observed before moving to the next step.

Summary

The first stage of improving performance is to judge whether it presently meets its expectations. When the process, product, service, or technology is underperforming, the team examines the reasons and causes of the underperformance. Brainstorming is an excellent tool for identifying potential causes. Rather than just list these, it is best to visualize these using a cause-and-effect diagram or a more sophisticated cause-and-effect matrix. Once identified, these causes drive the innovation project as the team attempts to eliminate or alleviate the influence of the causes.

Discussion Questions

1. Identify a problem where underperformance is an issue.
2. Map the process (Appendix B) or highlight critical process steps.
3. Identify causes of the underperformance by using a technique such as brainstorming.
   a. Once complete, use a fishbone diagram to identify cause and at least one effect.
   b. Use the cause-and-effect matrix on a series of causes (no more than five). Fill in the diagram (including requirements). What is the final score, the critical cause?

V—Validate with Measurement (Underperformance)

For this phase and outcome, the emphasis is on data collection and analysis. It is critical to validate the changes made that verify that performance now meets expectations without frequent upsets or changes. Key items to consider are

- Predictability
- Influence of lesser effects
- Control plans that ensure compliance

Validation (Figure 5.9) is evidence or proof needed to initiate the improvement. Once the team identifies the reason for the underperformance, then evidence (data) must determine the following:

- Its overall effect and influence
- The amount it contributes to consistency (predictable versus chaotic)

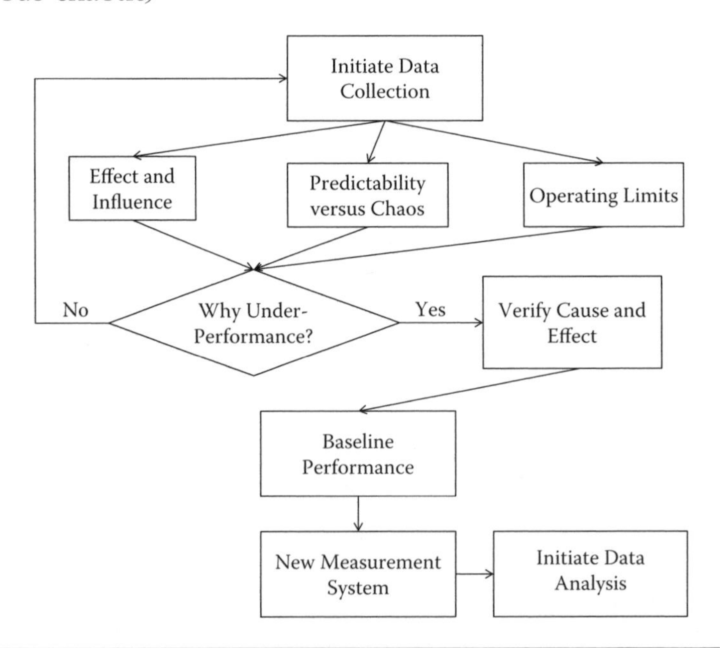

Figure 5.9 V—validate with measurement.

- Operating limits (settings and controls to maintain optimal performance)

This is probably the most underrated step in all the steps associated with improvement. First, determine the overall effect of the particular cause:

- Is the cause occurring erratically or chaotic (without notice or provocation)?
- Do specific situations (weather, personnel, and raw materials) result in changes that affect performance?
- Is the effect's influence consistent over time?

Causes or reasons that are consistently affecting performance are typical drivers of poor performance. These are easy to identify, and therefore relatively easy to fix. Erratic causes are much more difficult to find and much more difficult to fix. For these types of causes, there needs to be a constant monitoring of the process that produces the product, service, or technology. In addition, the influence of the cause or reason is also of concern. If the influences are weak but consistent, a series of applied controls can eliminate the effect. If the influence is moderate to strong, then specific actions are required to eliminate the effect, including the following:

1. Modification of the process (steps, activities)
2. Replacement of parts, resources, or personnel
3. Sophisticated controls (including modified action plans)
4. Change of procedures, tasks, and routine

It is critical to assess and understand both the effect of the cause and its overall influence.

Another concern is predictability of the cause and its influence on the effect. If a particular cause occurs predictably, then applying a set of controls will minimize its effect. If the cause is intermittent or erratic, Global Targeting recommends

Failure Modes and Effects Analysis (FMEA) Note: Failure = loss of performance

Process or Product Name			Prepared by:		Page ___ of ___
Responsible			FMEA Date (Orig) __________ (Rev) __________		

Process Step	Critical Element	Potential Failure Mode	Potential Failure Effects	SEV	Potential Causes	OCC	Current Controls	DET	RPN	EOC	Actions Recommended	Resp.	Actions Taken	SEV	OCC	DET	RPN
Identify Process Step (if needed)	What is the critical element or part?	**In what ways can this go wrong (fail)?**	What is the consequence on Performance?	How Severe is the failure effect to the project objective?	What causes the Critical element or part to go wrong?	How often does a cause occur?	What are the existing controls and procedures (inspection and test) that prevent loss of performance?	How well can you detect the cause?			What are the actions for reducing the occurrence of the Cause, or improving detection?	**Who is Responsible for the recommended action?**	What are the actions taken with the recalculated RPN? **Be sure to include completion month/year**				
									0								0
									0								0
									0								0
									0								0
									0								0
									0								0

Figure 5.10 FMEA template.

more and frequent monitoring of the overall process. One helpful tool is the failure mode and effects analysis (FMEA) (see Figure 5.10 and Appendix F), which examines possible failure modes (how and when an element will fail), consequences, and resulting actions. For those with a need for more and frequent process controls, use the control FMEA (Appendix G), which establishes action, contingency, and final control plans.

Whatever the outcome, it is important to validate with measurement all causes and their effects (influences) on performance.

One helpful tool for controlling the influence of causes, especially negative ones, on performance is the use of operating or specification limits. These limits should prevent a process or provider from creating an unacceptable item. These limits are normally set to control both acceptable items and erratic or uncontrollable changes.

Is there critical information unavailable for evaluating or monitoring performance? Would these presently unavailable data be able to diagnose problems or confirm deficiencies? For these situations, new measurements may be required to monitor the process to prevent failures. If no formal process management system exists, then create a set of measurements and measurement criteria to determine performance. Measurement systems can be as simple as a yes/no response up to and including sophisticated measurement equipment and software. Rather than purchasing or creating your own measurement system, it may be wise to collect data using a data collection checklist (Appendix I). Each time there is underperformance, indicate the occurrence with a check mark. These data provide the information about the process and the customer, but in a simplified manner. Whatever the choice, it is important to measure and evaluate the causes and reasons of a particular problem.

Summary

When underperformance occurs, it requires a thorough investigation of causes and reasons. This section discussed how to collect data to validate performance and identify critical causes that result in underperformance. The tools enable the innovation team to determine which causes affect performance and rank these according to importance. The highest-ranked causes are those deserving of the most attention. Those critical causes require repair, replacement, modification, or improvement to deliver expected performance. In addition, tools for controlling these causes enable performance to remain consistent.

Discussion Questions

Consider a product, process, or service that is underperforming. Detail the steps and possible tools your team would use to understand the reasons for the underperformance.

Consider the following. How will you:

- Identify the causes?
- Evaluate the causes?

- Measure the causes?
  - What instrument will you use?
  - How is it verified for consistency?
- Assign priority?
- Take actions to eliminate the reasons and causes?
- Establish a new baseline for performance?

Improving Performance to Exceed Expectations

O—Operational Profile

When trying to increase performance (beyond the expected baseline), Global Targeting recommends a slightly different strategy (see Figure 5.11). For this situation, improvement in performance occurs by identifying elements and components that may perform better than expected, thereby increasing performance. Rather than introducing a new item, use existing

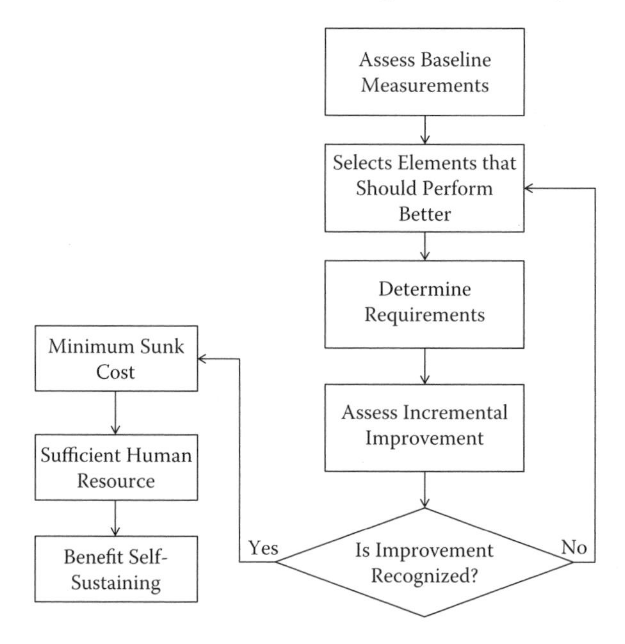

Figure 5.11 O—operational profile (overperformance).

elements or components as a method to increase performance. This stage requires modification or adjustment of (or replacing) a component or element when performance is recognized as less than desirable by the customer or user. For example, suppose you want to decrease the time it takes to get cash from an ATM machine. One way to do this would be to look at components or elements that make up the transaction or process when it is used is to initiate the transaction. For instance, if the customer requests $200 every other week, then it may be possible for the bank to recognize this recurring pattern and rather than going through a set of screens or menu item, the software could just offer this option to the customer after he or she properly identified himself or herself. This is an example of modifying the components or elements to increase performance. It may in fact be a very simple example, but it is representative of how to improve performance within a system that is already meeting its expectations. Consider a second similar example. What if the bank or financial organization permitted individuals to request funds from the bank on the Internet, and when they reached the ATM machine and signed in, they automatically received the funds. This again would be an example of how to increase performance by modifying components or elements that make up the particular process. We do admit this could be an everyday improvement, since most people would probably not consider a faster turnaround time at the ATM as being innovative. The example, though, does demonstrate our point that a need must be satisfied if this is to be considered innovation.

Identify the components that could modify or change the existing item resulting in improved performance. Assume these components are operating within limits but not at peak efficiency. Once identified, then:

- Assess baseline performance (elements operating within limits or expectation).

- Determine the elements to improve (modify or change) so that performance increases beyond the baseline. Remember, sustained performance is the goal. Determine the requirements needed to keep these components and elements at peak efficiency—or modify/change these to increase performance. This is not about significantly changing the process, but the elements (inputs) to the process.
  - Assess the incremental improvement and its effect on performance.
  - Can this incremental effect be recognized and sustained?
- Eliminate inconsistencies or chaotic episodes to possibly achieve a level of performance never before experienced.

Here again, data collection should help to identify these incremental changes. Of course, changing the process could be required, but the added expense may not be worth the benefit. Athletes do this all the time. They may train differently (process change), or they may change the exercises done (elements or inputs), but the result is to increase performance. For it to be innovative, recognition of the performance improvement by others is critical.

The outcome for this stage is truly an opportunity, and therefore requires the following assumptions:

- The investment (sunk cost) must be far less than the potential benefit.
- Resources (human) must be allocated to achieve and support the objective.
- Establish an improvement goal that represents a realistic increase in performance.
  - This tells you when to stop or reevaluate your efforts.
  - Be sure that the customer or user recognizes the improvement in performance and that it is self-sustaining.

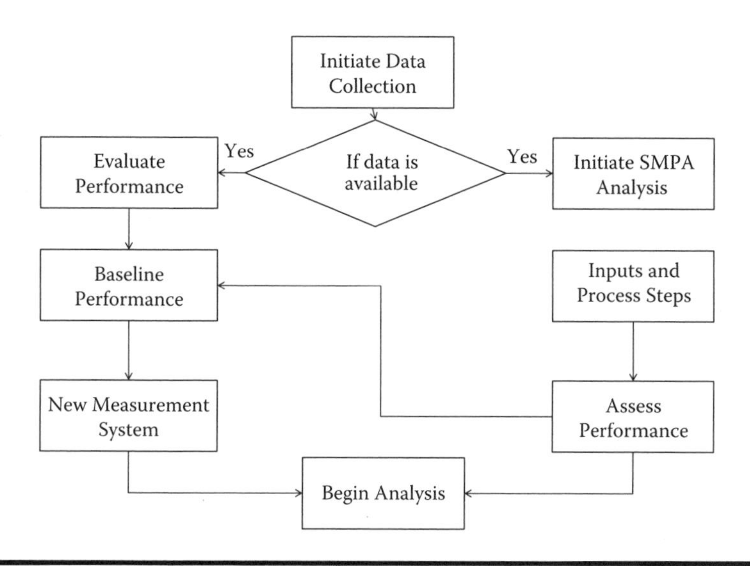

Figure 5.12 V—validate (overperformance).

V—Validate (Overperformance)

Validating with data is essential to identifying the components or elements selected for improvement. It is easy to find elements that can yield performance improvements, but for innovation purposes, the search is for significant improvements (Figure 5.12). A combination of improved elements that contribute to sustained new levels of performance is the most likely scenario. Data analysis can be timely but complex, and involve additional resources. One method to reduce the time to collect specific performance data is use of the success modes and performance analysis (SMPA) diagram. Figure 5.13 presents the chart that borrows and expands upon the FMEA template and reasoning. The SMPA analysis evaluates inputs (components or elements) that increase performance and maintain and control sustained performance. The SMPA tool permits the team to examine elements (components or parts) that may affect performance positively. Acknowledging that performance can increase, the opportunity lies in determining if the element is capable of sustaining performance over expectations. SMPA is a reasoning tool that highlights existing elements and the potential these

Success Modes and Performance Analysis (SMPA)

Process or Product			Prepared by:	Page ___ of ___
Team:			SMPA Date (Orig) ___ (Rev) ___	

Process Step	Key Process Input	Potential Success Modes	Potential Performance	PRB	Potential Causes	NEP	Current Actions or Controls	SUS	SPN	Actions Recommended	Resp.	Actions Taken	CON	OCC	DET	SPN
Process Step	What is the component, part, or element?	In what ways can the component, part, or element improve?	What is the affect on Performance?	Impact Probability - Rate the chance of continued improvement	What could cause the component, part, or element to affect performance negatively?	How frequently would a negtaive affect occur?	What actions (controls) are needed for this improvement to be sustained?	How well can the improvement sustain increased performance?	Success Priority Number	What are the actions requried for mainataining improved Performance?	Who is Respon-sible for the recom-mended action?	What are the completed actions taken? **Be sure to include completion month/year.**				
									0							0
									0							0
									0							0
									0							0
									0							0
									0							0
									0							0
									0							0
									0							0
									0							0
									0							0
									0							0
									0							0
									0							0
									0							0
									0							0

Figure 5.13 Success modes and performance analysis.

elements possess. After identifying an element, the team begins the process by completing the SMPA worksheet (Table 5.1).

The completed worksheet information transfers to the SMPA template (Figure 5.13). Team members use the rating scale in Appendix E to evaluate the potential and viability of the element to sustain success. Use the tool to initiate conversations to determine what elements deserve immediate attention. What the tool lacks in scientific accuracy it more than makes up for in ease of use. Its purpose is to discuss and assess the efficacy of each element. Further evaluations should address topics such as maintainability and consistency (reliability). A cost-benefit analysis should follow to determine whether to consider or reject the element. SMPA focuses on the positive (unlike the FMEA, which uses similar reasoning) recurring influence on performance. Metrics (measurements) and a final decision can follow the use of this tool.

This is an excellent tool for identifying the elements that affect performance as well as establishing methods of control or monitoring. You can find directions in Appendix E.

Table 5.1 SMPA Worksheet

Input	*Potential Success Mode*	*Potential Performance*	*Potential Causes*	*Current Controls*
What are the elements or components that directly affect performance?	How can these elements (or inputs) improve?	What is the effect on performance?	What could cause the component, part, or element to affect performance negatively?	What actions (controls) are needed for this improvement to be sustained?

Summary

It is noteworthy that differences between the two improvement outcomes are significant. When improving performance through cause-and-effect analysis, the objective is to identify the cause of reduced performance, validate the effect, and provide a fix. When attempting to improve performance beyond present expectations, the focus is on improving elements or components that directly affect performance. After identifying and validating these elements, the process of improving performance can commence.

Discussion Questions

1. Consider a performance measure of a process, service, product, or technology that can exceed present expectations.
   a. Complete the SMPA worksheet only.
   b. Is there a certain success mode that would ensure that performance could be consistently improved?
   c. What causes the success mode to remain in accelerated performance mode?
   d. What can be done to consistently maintain the performance level?

A—Analyze and Align (Underperformance)

In a traditional problem-solving methodology like Six Sigma, the analysis phase consists of a number of statistical tools and techniques. Although statistical analysis can provide an excellent evaluation of small differences, it is not required when looking for large differences. That is, statistical analysis is useful when the influence of causes and reasons is small but pervasive. A statistical approach does provide for understanding concepts such as dependence, interaction, and variability. All of these may be critical for improvements throughout the organization. Remember that the improvements discussed in this book are those perceived as innovative, with necessarily large changes in performance. Small differences are not of great importance when determining whether an improvement is truly innovative. Therefore, we will not make use of sophisticated statistical analysis for an improvement initiative. This is not to say that statistical analysis is unimportant. It is very important in distinguishing differences on a day-to-day basis. Innovation, however, is not incremental improvement but improvement that demonstrates a marked increase in performance. There are, however, many tools that are applicable to both situations. Descriptive data analysis with charts, graphs, and tables is very useful to determine differences and identify possible causes for the loss of performance. Examining the analysis consists of understanding the overall pattern and not reacting to individual high and low points. Concern yourself with looking for patterns and trends and then reacting to unusually high or low points, referred to as outliers. Many of the practices and principles developed during problem-solving training are useful for innovation. However, interpreting the data differs from standard problem-solving strategies. It is the process or elements resulting in large differences or changes that may in fact be the best information for those working on an innovation project. The real mindset is to move away from fixing a problem to increasing performance (Figure 5.14).

In fact, analysis is truly more than just statistical testing. Analysis involves understanding, reasoning, and judgment regarding the causes of underperformance in the influence of its effects. You can assume that when an item requires improvement from an innovation perspective, it is more effort than just fixing a problem. Innovation requires a different process when thinking of the existing process and its contributions to performance. This is what makes innovation, for some, daunting and seemingly impossible. In essence, it is more of a mindset than a process reset. The mindset surrounding innovation is that customer or user needs are the

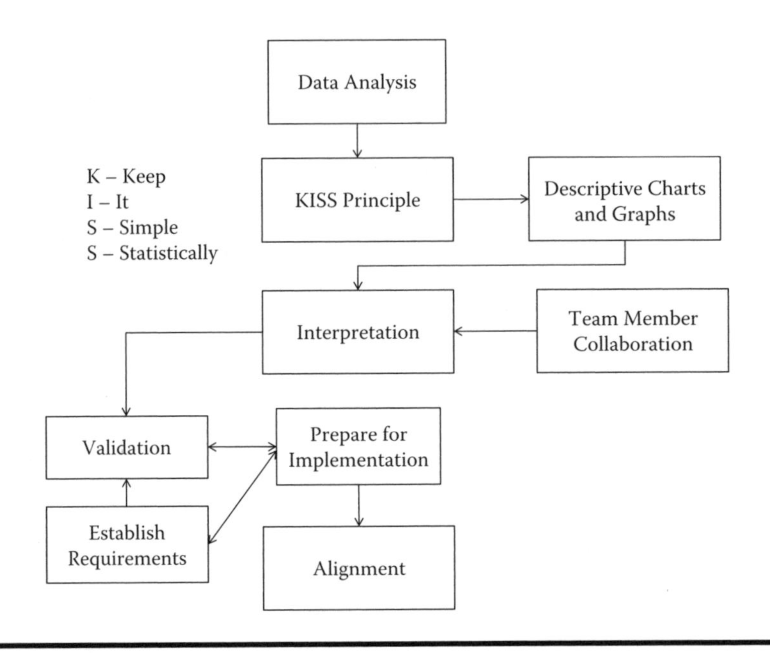

Figure 5.14 A—analyze and align (underperformance).

drivers of purchase behavior, differentiated by the performance the item is perceived to have (or not have).

One concern companies and organizations will have is being able to separate day-to-day improvements from innovation efforts. Applying a problem-solving methodology to a process is extremely good business sense, but it is not innovation. Therefore, the tools for innovation will be different, as the results will be sufficiently greater than what would be achieved by incremental improvement.

Interpretation is critical in the analysis phase. When interpreting any data, be sure to validate any suspicious results. Checking the result further validates the data value. If there is an obvious error, then remove the data point, as this adds validation to the overall data set. As part of the improvement, implementing new requirements may be required. Validation of these requirements is a must to remain consistent and unbiased.

Alignment, for the underperformance step, is a combination of validation and analysis (Figure 5.15). Analysis requires both measurement and interpretation. Aligning both measurements and

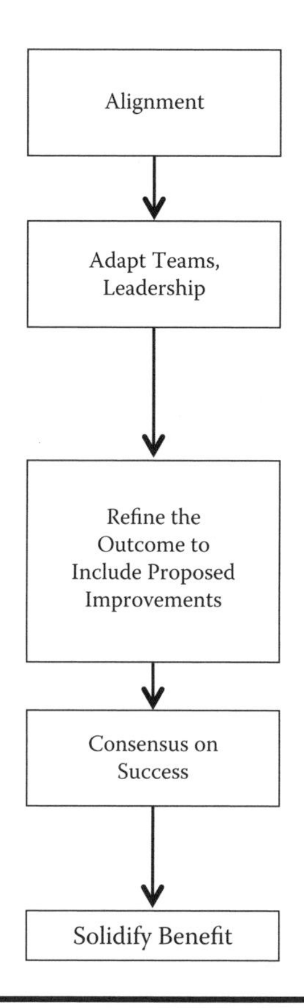

Figure 5.15 A—alignment (underperformance).

causes for improvement can provide the strategic direction for this stage of the process. Alignment includes measuring performance and linking this to changes (both positive and negative) to those elements that require or need improvement. However, analysis alone does not provide the full set of information regarding the data collected. Interpretation of the data requires experience and knowledge of the process, the needs, and the purchase behaviors of the customers and users, as well as an understanding of past performance. This is critical for innovation since the intent is not just to improve the item but to improve the item in such a way that it is recognized as innovative.

Improvements will become sustainable when those items that affect performance are understood and controlled (monitored). In understanding causes or reasons for less performance, it is necessary to understand those components that influence the process as well. That is, take a holistic view of the process as it pertains to its performance. This is a deviation from the traditional improvement projects that focus on single or multiple components of a process to alleviate a problem. This holistic perspective considers that all components operate in tandem with each other, affecting overall performance. Beware of a myopic focus, but remember the individual contribution of any single element, component, cause, or reason.

When improving performance beyond current expectations, remember the need to align people, process, procedures, and resources. Given the nature of changing the existing paradigm, regarding performance, align team members and leadership to the expected outcome. Refine what will become the new requirement (new baseline), including those elements that are responsible for the improvement in performance. Finally, achieve consensus on what to expect and what defines the benefit objective.

Summary

Analysis and alignment are critical for both innovation improvements. Before committing to a modification, the evidence (data) collected and the consensus should agree that initiating the improvement will meets its objective. This stage (or phase) may be the most important step since it confirms the results obtained to date. The team has identified the modifications and agreed that they will benefit the organization and be sufficiently successful for consumers.

Mini-Project

Create a set of decision rules for validating and aligning a process. Include the following:

- Data signals to clearly indicate the process is changing
- Early warning signs that the process may change
- Rules for correcting or changing the process
- Rules for shutdown or contacting the appropriate person in charge

A—Adapt (Overperformance)

Adapting to any changes suggested by the analysis should be simple for the team after its involvement with SMPA analysis. This is a great alignment tool for the team. Adaptation may be more organization-wide as changes are made to increase performance. The team must consider the stakes of those that created the original process. How can this innovation team come along and suggest improvements to a process that is already meeting expectations? Management leadership and communication are critical elements for success. Again, applying performance-enhancing criteria to a process that produces erratic performance is foolish. The team will be trying to improve something that is inconsistent, by nature. This result is frustration and lost opportunities. Adaptation is more an external outreach and requires more than just communicating positive results. It requires leadership to lead the changes made and to support the team's suggestions and progress (Figure 5.16).

Often successes encounter resistance. This is typical when a process, product, technology, or service needs improvement. It is typical to experience resistance even when the evidence proves a new fact. If a group feels that the change will cause them additional work, job disruptions, etc., there will be resistance. A strategy to address resistance is communication and involvement. Once employees and managers experience the benefits, resistance will gradually reduce. Support for the project may also suffer over time as leaders and managers focus on the latest problems. Communication and involvement are critical to keeping an interest level high. Involving

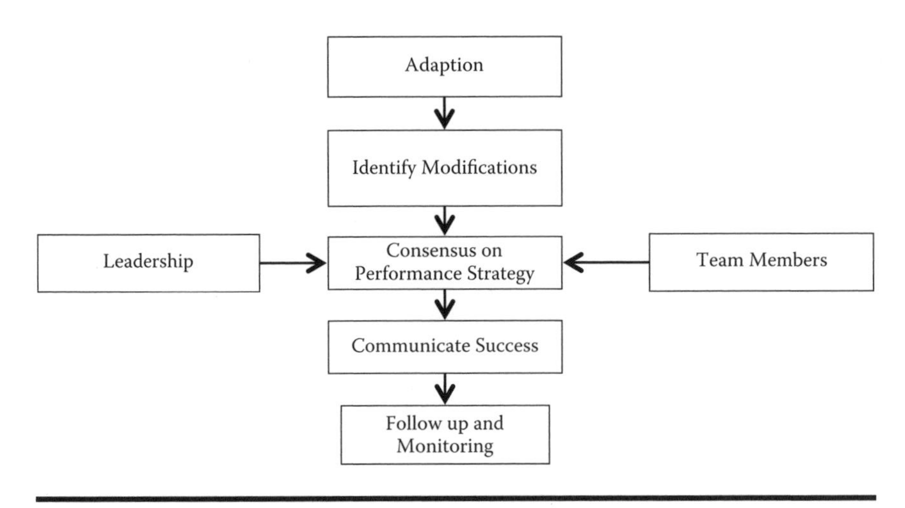

Figure 5.16 Adaptation process step (overperformance).

management with the team keeps the individuals connected to the project.

As with ENOVALE steps, communication and alignment are critical for success. Additionally, monitoring of the changes provides a method to both evaluate and react to changes, thus decreasing the chance of reduced performance. It is best to monitor all causes to keep performance from exceeding previous expectations. Customers and users will react to the improved performance and create their own, new baselines, so the provider must increase its efforts to monitor performance and customer/user needs.

Summary

When confirming decisions to modify, change, or replace an item, the team, leadership, and customer must adapt (or abandon). Adaptation is an external outreach to the organization and user. The purpose is to prepare the customer for the improved performance, calibrating the baselines to the new performance expectations. At this stage, external and internal communications must focus on how this new level of performance exceeds expectations. Leadership and management must be onboard with the changes, supporting the need for enhanced performance. Adapting the customer/user is the final element for success.

Discussion Questions

1. Identify how to adapt the following to introduce enhanced (over)performance
   a. Communications to and with the customer/user
   b. Internal operations
   c. Leadership and managerial practices
   d. External resources and suppliers
2. Discuss the barriers and strategies to overcome:
   a. Complacency with the norm
   b. Resistance to change
   c. Leadership disinterest

L—Link to Improvement/Performance

This step describes both under- and overperformance improvement strategies (Figure 5.17). Incrementally the improvement should yield a benefit to the organization as much as be seen as innovative. Therefore, does the benefit register as either a cost-saving or a profit enhancement opportunity? Could it open new business opportunities or refresh existing businesses? At this nearly final stage, it is important to gauge success. If it is not a financial benefit, is the benefit recognizable and ongoing?

It is difficult to identify and "book" benefits to an organization or business. Many current accounting systems cannot identify specific benefits associated with process improvements. Costs go up and down at such a pace that improvements are often lost and then said to be nonexistent. For example, if a department saves $100K in benefits but monthly costs can vary by $200K, how can the accounting system identify the benefit? Well, in plain English, it cannot. An opportunity with a true improvement is confounded with

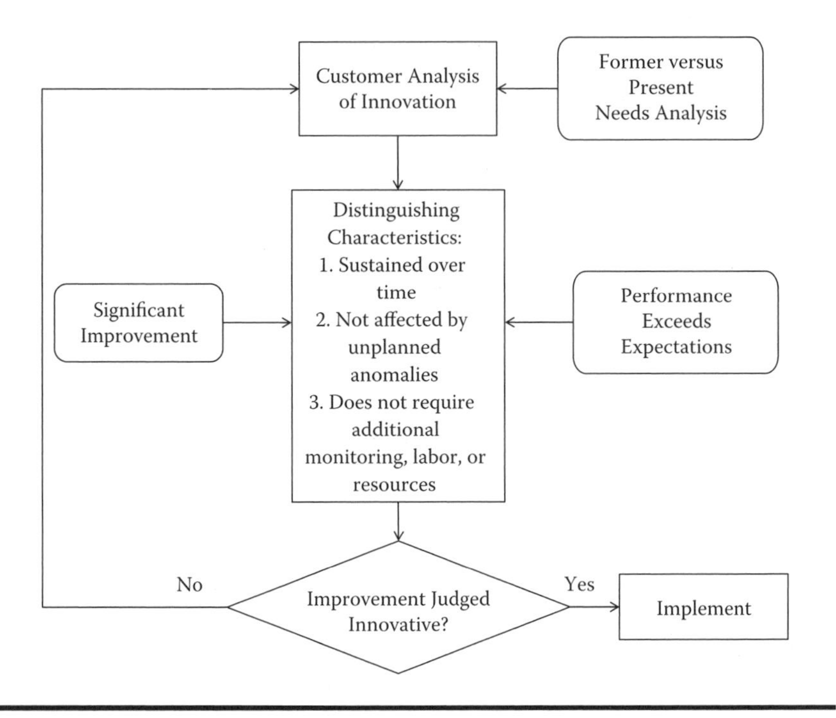

Figure 5.17 L—link to improvement/performance process step.

other, nonrelated costs, often causing a drop in performance and financial gain. The difficulty lies in the accounting system that lacks definitive detail or uses average costs that "smooth out" variations. This is why we suggest you look for confirmation from the customer or user. Only then is the innovation verified. The difference in this step, from previous ENOVALE steps, is that the focus is not on costs, profits, revenues, etc. Performance is the only interest, and whether an increase in performance will in fact be judged innovative by the customers and users. For this step, we link improvements back to performance. That is, a change in improvement should yield a change in performance. Consider the following:

- If the performance is unsustainable over time, then the improvement is not innovative.
- Do unplanned anomalies (erratic behavior) interfere with performance?

- What is the effect of any negative consequences?
- Does maintaining the improvement require additional monitoring, labor, or resources? Is it a cost issue to maintain the innovation?
- Finally, is there consensus among the team on the project's failure or success?

Innovative outcomes will definitely provide a benefit or benefits to the organization. Clearly capture these at this stage. Internal benefits such as cost reduction or profit enhancement are not something that the public may see or experience. Internal concerns are quite important, but again, do not on their own affect or influence innovation. Most small or medium-sized company financial systems are incapable of tracking improvements in cost reduction or revenue generation to the project level—generally, this is only at the department level. Therefore, the innovation project should focus not only on internal benefits but also on meeting or exceeding external needs. Be aware of the concern that the innovation opportunity is relatively independent of other effects, so as not to lose the benefit. Monitor the performance of the innovation; financial gain will come from increased levels of competitive advantage.

Summary

The key takeaway from this section is to monitor performance to determine overall improvement. Since the entire improve process is built upon under- or overachievement (in performance), it is critical to thoroughly measure and evaluate performance. This is not to negate internal benefits derived from innovation. Performance is the driving metric upon which many decisions are made and the innovation is judged. Traditional measures, such as costs and benefits, are realizable at this stage. The wise leaders or manager will understand the reasons for driving performance directly relate to innovation perceptions. If a producer can say that its product is improved, then customers/users expect a relative improvement. If a producer says that the improvement is innovative, the customer must experience either an existing need or a new need, and that performance is well above previous expectations.

Discussion Questions

1. Identify three or more benefits that a customer or user could experience regarding an innovative improvement?
   a. How do these relate to one another?
2. Devise a strategy for meeting these benefits (detailed in question 1).
3. Can this strategy be implemented at the L stage of the ENOVALE strategies process?

E—Execute Controls (Underperformance)

Unlike other ENOVALE strategies, this step requires that specific controls be established to maintain performance. A control is simply a method of monitoring performance, and an action plan if the items falls outside of a specified requirement or limit. Controls provide a game plan for a process to maintain its expected objective. Controls can be either procedural or automatic, but they must be more than pure guidelines. Procedural controls are those that come into existence when the process begins to perform below expectations. These are related to changes in the process that are different from the day-to-day set of routines and practices. Backup plans are excellent examples of procedural controls (Figure 5.18).

Automatic controls are those built in to the system to react when the performance is below expectations. Automatic controls may be machine or process driven with a set of actions plans put into place, by people, to correct the situation affecting performance. A good example for automated controls

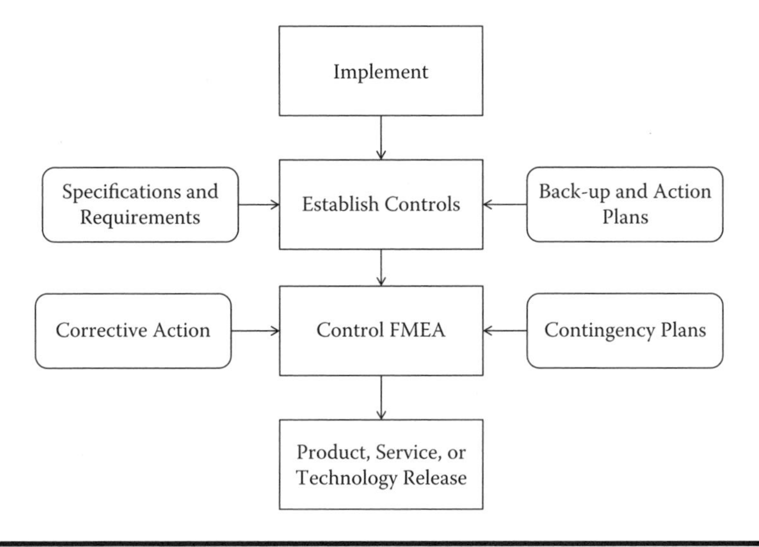

Figure 5.18 E—execute controls.

would be a set of fire doors that would close when a first signal or alarm is initiated.

An action plan is appropriate when processes involve procedures that must occur in order to bring the item back into the state of control. It is a set of procedures to initiate and return a process to expected performance. Action plans are not an immediate response, as would be an automated or procedural control. These are longer-term sets of actions taken to readjust a process so that its performance returns to expected levels.

One exceptional tool that can be used for both an analysis and a control purpose is failure mode and effects analysis (FMEA). It serves the purpose of both examining the reasons why performance was below expectations and providing a plan to prevent this action from occurring. In addition, it provides for an assessment of risks and consequences as they relate to the actions put into place to prevent performance from dropping below expectations. It has the capability of identifying probable cause for less than expected performance. It has the power of a cause-and-effect tool, and assigns risk to the actions taken and consequences of those actions. It is also useful for monitoring which component or element is in most

Failure Modes and Effects Analysis (FMEA) - Control Process

Innovation Team project #		FMEA Date:	FMEA Revison #

Control Process													
FMEA No.	**Process Step**	**Improve Elements**	**Control Elements**	**Failure Modes**	**Severity**	**Potential Causes**	**Occurrence**	**Contingency Plans**	**Responsibilities**	**Remedial Action Plan**	**Detection**	**R P N**	**Audit Item**
FMEA No **(Tracking only)**	What is the process Step being Improved?	Which critical elements were selected for Improvement ?	What actions will be taken to control each Improvement Element?	In what ways can the Control Element fail?	How severe is the Failure Mode to meeting project objectives?	What causes the Control Element to fail?	How often does the Cause occur?	What are the plans to prevent the Failure Mode or actions to take when failure occurs?	For each Failure Mode, list who, what, when, and how the remediation will occur	What are the checks and balances? **(Document the Remedial Action Plan)**	Will the checks and balances be able to detect non-compliance?	What is the Risk Priority Number?	Yes/No? **(A high RPN requires an Audit Plan)**
												0	
												0	
												0	
												0	

Figure 5.19 Control FMEA.

need of control, that is, which element or component most influences changes in performance. It provides for an action plan with detailed instructions enabling a return to performance at its expected level.

Rather than using the standard FMEA, we apply a hybrid of this tool, called the control FMEA (Figure 5.19). Instructions for using this tool can be found in Appendix G.

Summary

Of all the ENOVALE strategies, this improvement strategy is aligned with traditional project improvement efforts. Many of the tools and techniques are similar to those used with other improvement projects. The major difference is that these techniques, when used properly, lead to innovative improvements. For improvement to be innovative, individuals must recognize that performance is greatly increased without sacrificing any of the needs that the item was providing. The increased performance should be enough to attain some competitive advantage. There will be a tendency to confuse this form of innovation with standard improvement technologies presently in use. Yet, improvement is often the most widely recognized form of innovation. Measuring, identifying, and controlling performance is the best method of maintaining the improvement. There will be a time when performance needs to exceed beyond present requirements, that customers/users adjust their expectations to consider this performance as expected. Customers/user will then wait for the next innovation.

E—Evaluate and Review (Overperformance)

As with all final sections or steps, this one involves putting into practice what was learned from the first six steps (Figure 5.20). All that remains is to implement the improvements and monitor the performance (from an internal perspective). Externally, monitoring customer reaction, demand, and perceptions (attitudes) of satisfaction is critical. When implementing any innovation, customer and user needs must be carefully monitored and evaluated. Track the three dimensions of external (customer) as follows:

- Enhancement of existing needs and meeting unfulfilled needs
- Continued, extended, or new levels of performance
- Increased experience and knowledge of product, service, or technology

Additionally, track the three dimensions of internal (organizational) needs:

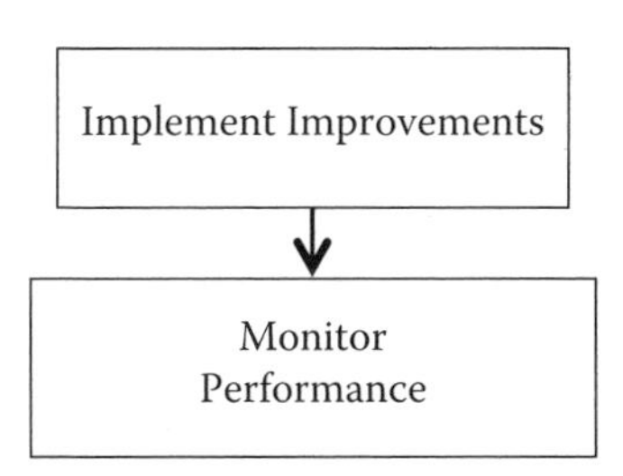

Figure 5.20 Evaluate and review (overperformance).

- Capability—performance, profit, competitive advantage
- Viability—immediacy, availability, high success rate
- Sustainability—life cycle, extended profits, and competitive advantage

Tracking customer/user needs keeps the business informed on what and when to release the next innovation. Those organizations that can quickly implement and satisfy needs will succeed. This information provides the impetus to begin an innovation project. Track needs to determine if improvement is required.

Once implementation and rollout are complete, be prepared to review performance results. Improvements definitely have a time limit—they are time-sensitive. However, products such as Tide detergent have survived for years by announcing that the product is "new and improved." This is why we suggest applying resources to those items that generate the largest prevailing need (and most stable needs). Otherwise, continue improvement efforts on a daily basis.

Summary

Implementation follows the validation of improved performance. This final phase or step is critical since the monitoring of needs generates the next innovation opportunity. Monitoring the needs (as well as performance) provides competitive advantage. Without this phase, it is impossible to know where, how, and when the next innovation opportunity is present.

Mini-Project

Underperformance

Construct a control FMEA worksheet for an existing process:

- Be sure to identify primary causes for change.
- Evaluate risk based on elements such as:
  - Timeliness to react
  - Impact (influence) on all process elements
  - Severity of causes
  - Failure effects and consequences
  - Reaction modes
- Prepare a simple set of action items that will prevent future failures.

Overperformance

1. Construct a set of guidelines (timelines) for implementation.
2. What barriers may prevent implementation from proceeding smoothly?
3. Once implemented, name at least one characteristic that needs to be carefully controlled?

Chapter 6

Innovative Change

Introduction

Change is pervasive in organizations today. Change, from Global Targeting's perspective, is that which produces positive results—innovative change. With respect to traditional ideas of innovation, change represents the most striking divergence, since it is a transformation of the decision-making process rather than more tangible items such as product and technology. Someone must initiate, administer, and carry out the change. Therefore, change has a very human or individual characteristic. This fits naturally into innovation, since innovation begins and ends with the individual (Figure 6.1).

Because the concept of change involves many aspects, the definition of change, used in this context, is that of replacing what exists with something judged better (positive). This positive nature of change is innovative since it must meet new or existing needs and the performance meets or exceeds expectations. Change follows the basic premise of new or improved.

This chapter will detail the ENOVALE strategy for innovative change. Remember that this strategy requires that ENOVALE solutions precede the application of this strategy; otherwise, the process may produce less than acceptable results.

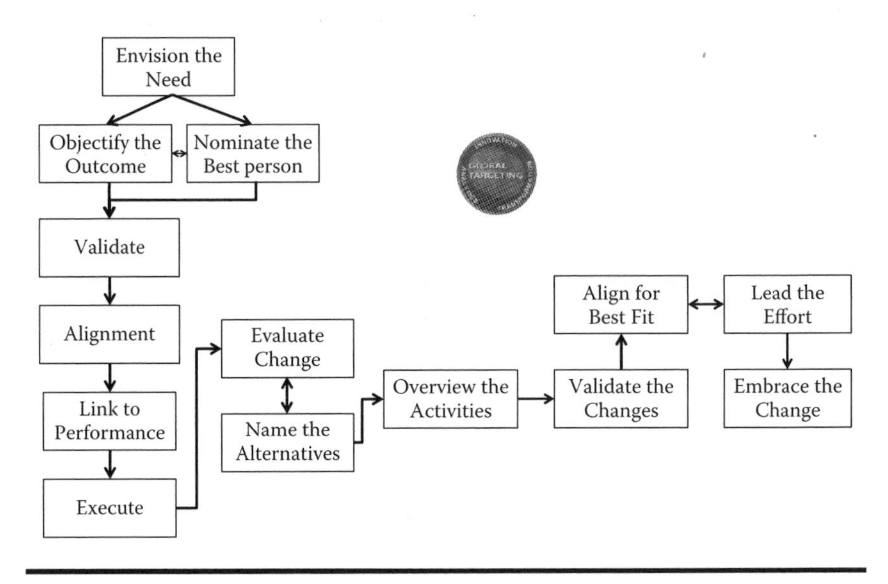

Figure 6.1 ENOVALE solutions and change strategies.

E—Evaluate the Alternatives

Change is never easy, but it is inevitable. Fear often accompanies change. Many people have come to view any change as something that negatively influences their lives. In organizations, whether for profit or not for profit, change is inevitable. People resist change, accept it, or are ambivalent. The fact that individuals react to change is fundamental to understanding change from the perspective of innovation. This reaction will relate directly to consumer behavior regarding use or purchase of a product, technology, or service. The change, which becomes the transformation in the outcome, will be positive, negative, or neutral. The largest difference between innovative change and new and improved innovations is the focus primarily on individuals. Measuring the outcome of change involves both a tangible and an intangible response. Change can be measured tangibly by examining the results or intangibly by capturing the attitudes of those individuals most affected by the change. As with all innovation, the transformational process guarantees success or failure. Change

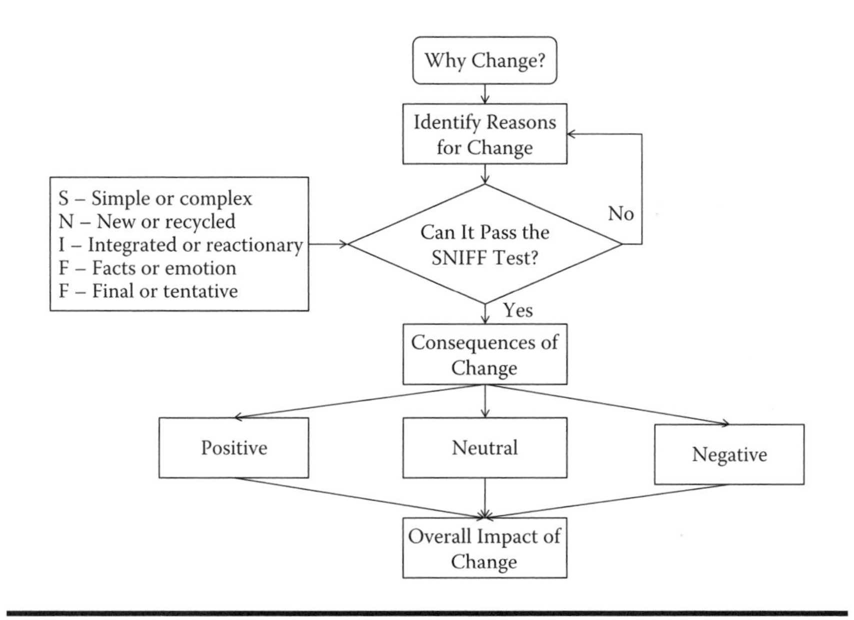

Figure 6.2 E—evaluate the alternatives.

creates a positive, neutral, or negative result. Obviously, only positive results will be the most beneficial for an organization (Figure 6.2).

Before change begins, there must be a deliberate and pervasive reason for the change. Identify the reasons for the proposed change. Can the proposed change pass the SNIFF test?

- S: Is the need for change simple or complex?
- N: Is there a new reason for the change or simply a recycled idea?
- I: Could the change be judged as integrated (for a specific goal) or reactionary?
- F: Is the reason for the change based on facts (evidence) or emotion?
- F: Will it be a final (permanent) change or one that is temporary?

If there is significant disagreement between responses for the SNIFF test, then reconsider the objective of the change. This simple exercise provides an opportunity to reach quick consensus.

Given the human dynamics of innovative change, the focus of this process will be on the human aspects of change. Change affects humans in the following ways:

- People are replaced, rearranged, or reassigned.
- It creates a distinctive aftermath—a new environment in which individuals must adjust and perform within.
- It affects individuals beyond basic needs—individuals are affected emotionally and psychologically by change.
- Change disrupts the status quo; it brings about a new order. Sometimes this order is efficient and effective, and other times it is negative and debilitating.

Change affects people differently from different cultures. Think about the very negative effects on the native (indigenous) peoples that lived in North and South America prior to the arrival of the Europeans. Their lives, livelihood, and very existence became dependent upon the Europeans. Exploration was a noble act (and an economic reality) for these Europeans, and in fact was a result of a change in thinking that occurred during the Renaissance. What was considered progressive in the European mind was considered destructive and devastating in the native people's mind. This is why we talk about change as it relates to the consequences of the change. For innovation purposes, change must have a positive outcome and satisfy needs that are not being fully satisfied. Alignment is critical for success. It is similar to the concept discussed in previous chapters.

The consequence of change is a consistent theme throughout this chapter. Every change will produce a negative, neutral, or positive outcome. Negative outcomes are greatly reduced by understanding, planning, and adjusting for these consequences. Recall the scientific (Newtonian) principle that "for every action, there is a reaction." With change, the same is true. At times, consequences will appear to be an insurmountable barrier. However, by acknowledging these consequences, understanding their effects, and devising a method to avoid (compensate) for as many as possible, the result will ensure a positive or neutral outcome.

As discussed in previous paragraphs, change that occurs is transformative. What initiates the change, however, is a decision. Therefore, the decision-making process will be critical to ultimately enacting the change. Issues such as leadership, advocacy, and good project management are all critical to achieving a successful outcome. Traditionally, most decisions regarding innovation projects were initiated at the executive level in private, without the assistance of those directly affected by change. This process works quite well for day-to-day decisions that do not require extensive evaluations or major changes in operations. The process, however, will not work as well when dealing with an innovation project. One-way decisions, generally made with judgment, emotion, and experience, are insufficient for innovation. Prevailing wisdom, such as last-in first-out (LIFO), is common. Because decisions are experience and opinion driven, resistance can be rampant. It is the process of making the decision that results in the change made and the resultant positive or negative environment.

Many have experienced the problems and challenges brought about by change. Some of the negative implications of change are

- Loss of productivity
- Increased costs
- Decrease in motivation and morale
- Need for emotional and psychological counseling
- Initiation of negative strategies used by employees to protect themselves
- Rise of employee resistance
- Risks are seen as harmful and thus avoided when possible

Negative implications create a negative environment harmful to employees, management, and customers. So the question that comes up is: How does innovation survive in a negative environment? Given its organic nature, it does not flourish well in this environment. Yet, the environment can change, permitting innovations to flourish. This is the precise reason for implementing the ENOVALE solutions approach.

Not every change will result in a positive outcome. This is the reason for the ENOVALE strategies—to ensure that change is perceived as innovative (at best) and necessary (at worse). Change is often the most dramatic innovative effort an organization can undertake. Its effects are felt on a daily basis, and it demonstrates the dynamics (effectiveness) of organizational decision making—it is real time! Now, some may think that innovative change is more internal innovation, given the discussion of employees and changes within the environment. The same is true, and often experienced, by your customers and users. Many changes may affect them at the expense of losing them to a competitor. Consider the example of an organization that reduces its workforce to eliminate excessive costs. Fewer workers may easily translate into more complaints by customers and users. The small savings attained by the workforce reduction may in fact relate to a tremendous loss of profitability.

Customers and employees will recognize a significant positive result as innovation. Innovative change needs to be proactive, for the customer, the employee, and the overall organization. It must enhance productivity and effectiveness. Innovative change must occur in a form of cooperation and positive and truthful communication. It must produce improved results (improved performance) and be recognized by the customer (and employee) as innovative (meeting or exceeding needs). That is, the change must satisfy a need that is not being satisfied with other means. Finally, customers, users, and employees must perceive the change as necessary and beneficial.

Summary

Before considering a major change, evaluate its overall effect. Identify significant reasons for the change. If the reasons are compelling, then continue on; if not, consider other approaches. Use the SNIFF test to determine if the proposed change moves forward. Consider the consequences of change and its overall effects on the individual. Even a neutral effect (impact) can positively affect the individual when the outcome yields benefits. Innovative change is recognizing that improvement occurred. It may be some time before individuals call this innovation, but the reality of a significant improvement will drive positive results for the organization.

Discussion Questions

Consider a recent or planned change in an organization. Assess the change using the SNIFF test components.

- Did the results change the outcome? Would you institute the change?
- Did any element change during or after the experience?
- Would you use this simple tool going forward to quickly evaluate change?

Prepare a set of statements for the following:

- Identify a recent change that was made in either your organization or your life.
- Identify one potential alternative.
- Using the criteria described in this section, evaluate the alternatives.

- Did you or your organization make the right choice?
- What would you do differently next time?

N—Name the Alternatives

The first step overviews the entire process, using the results of the SNIFF test. Every change brings about consequences. The consequences of change could go from dramatic to simple and be regarded (received) as either negative or positive. Therefore, begin with the result (the outcome), consider the obvious consequences, and evaluate the worth of the project. If the worth is valuable, potential exists for an eventual innovation opportunity. Assessing the overall value of change is critical before moving to this step. The second step reviews the alternatives and consequences (Figure 6.3).

Alternative thinking helps to visualize the entire planned change, and especially all of the consequences. Alternatives are viable approaches other than what is presently in place.

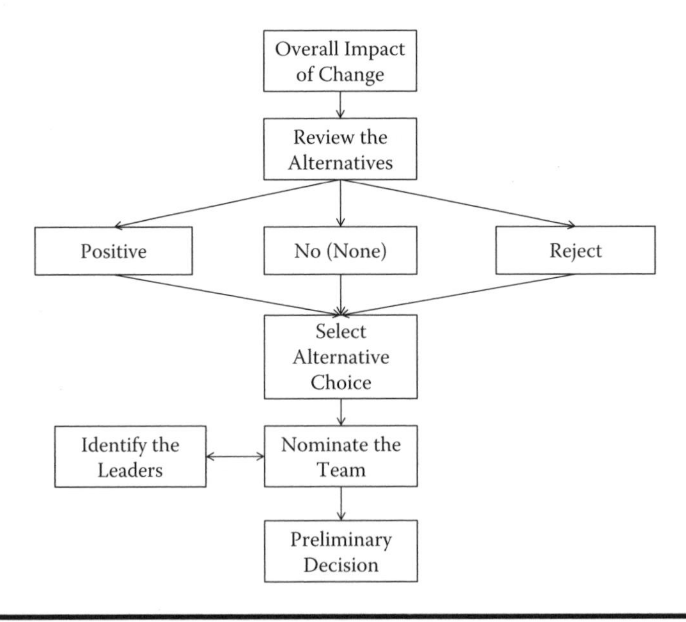

Figure 6.3 N—name the alternatives.

The intent is to reach the same goal, and one that reduces confusion, errors, or mismatched planning. Alternatives could be as simple as a workaround to a major replacement of various parts of the process. Alternatives provide an opportunity to view results by selecting different methods in which to initiate the change. Many times the innovation is found in the quality and effectiveness of the alternative. Alternatives should have specific features, such as

- Be 100% viable
- Achieve the same goal or an acceptable trade-off
- Meet corporate or organizational values, ethics, and goals
- Cause the least amount of disruption
- Address the welfare of employees; limit sacrifices
- Be accepted by customers or users (meet their needs)
- Not result in significant loss of productivity

With change, alternatives provide opportunities. When deciding to change, consider alternatives known to eliminate or reduce the effect of negative consequences. Viable alternatives provide the organization with choices.

Generating alternatives, however, requires a set of guidelines or criteria. Of most importance is the ability to rate or rank these alternatives in terms of their effect on employees, the organization, and customers/users. Therefore, when initiating a change, leadership, management, and teams should examine the positives and negatives associated with the decision. Then decide upon the alternatives. To determine which alternative is a best fit, assign an importance scale to the alternatives. To rate an alternative consider the following:

- Longevity (the time frame in which the alternative will exist)
- Comprehensiveness (capability of the alternative to function properly)
- Effectiveness (will the alternative significantly prevent mistakes, errors, inconsistencies?)

- Benefit (value of the benefit over time)
- Viability (how well will the alternative work?)

Choose the best alternative, even if it not ranked first. Consider the use of simulations or risk analysis to test the overall effectiveness of the alternative.

What if there is no alternative? Then the team must use what is available to accomplish the goal. Leadership and team members learn to adapt to the situation at hand. Without an alternative, the organization must accept existing conditions. One workaround is to measure, evaluate, and control existing conditions.

When alternatives are not possible, consider the following situations that require a decision:

- If a negative decision is the only alternative, then choose the alternative with the fewest consequences.
- People deal with traumatic change their entire lives and deal with the consequences as well.
- Do not underestimate people's ability to adjust to change and find a benefit that is positive from the decision.

Consider that few or no alternatives may suggest that replacement is preferable. Replacement is a decision to scrap the old system for something new (does not yet exist at the organization). Consider the example of the present IRS tax system (code). Some argue for use of a "flat tax," as changes to the present IRS system seem to increase time and confusion to complete a form. The flat tax would replace the present complex and confusing system. Replacement may involve people, procedure, policy, or process. For many this is an alternative strategy and one that deserves serious attention. If the outcome is positive, then replacement is a viable change strategy.

Finally, selecting the team is critical for initiating the project. Those individuals that favor innovative change should be the rational choice for team selection. This may fly in the face of prevailing wisdom that says that managers make the best decisions in selecting and assigning team members. In selecting the best individuals, there needs to be an understanding of who these people are and their overall contribution. First, consider employees who are mature, decision-making adults; they make decisions daily. Second, managers need to trust that employees will make an informed decision. Third, appropriate communication skills and technology must keep the manager and employees informed. Remember, when someone is trusted, that person is also valued. Research confirms that when employees are valued, they are more productive.

Part of the selection process involves who will lead the project team. Management's responsibility is to lead the change effort. The team implements the ENOVALE process (for innovative change). Use the ENOVALE process on a change with a proposed neutral or positive outcome. Use the human resources (HR) group to help guide the project, as this group (function) deals with change every day. Give HR the responsibility for finding and developing employees with skills in change management. Select team members with a positive attitude toward change.

Summary

Identifying and processing the alternatives prepares the team to begin the change management process. Alternatives must meet corporate or organizational values and objectives. Rating alternatives permits a better evaluation of their efficacy. If possible, use simulations or risk analysis to determine alternative potential. Finally, identify team members and a leader with a strong desire to recognize change.

Discussion Questions

1. Consider a change within your organization. What if the decision makers had chosen a different alternative?

  - What would the outcome be today?
  - Would it be positive, neutral, or negative?
  - If the alternative had been implemented, what would be its consequences?
  - Do you believe that the decision makers truly considered this alternative?
2. What should be the team selection process for identifying change-driven innovation?
   - Should it be experienced or empirically based?
   - Should the leader be named first, or after the team is assembled?
   - Are certain individuals better matched for this team?

O—Overview the Activities

Once the initial change process is reviewed, and alternatives added, consider the overall implications (Figure 6.4). In reality, every decision has multiple endpoints. Those decisions that most affect employees and customers have the greatest potential to be innovative. Consider a different approach. Think not only of the consequences, but also the repercussions. All change brings about repercussions. A repercussion is the aftermath effect (resulting effect) on an individual or group of individuals. It is natural to evaluate alternatives and their repercussions. As this can be a time-consuming process, consider only those alternatives that the team is seriously considering. The alternative is the activity, the repercussion, or the resultant effect. Assign risk to each repercussion, and for those that score high, develop a contingency plan to modify their overall influence. Use the Alternative and Repercussion Effects Analysis (AREA) (similar to FMEA), as exhibited in Figure 6.5 and Appendix H. This tool will help analyze the risks associated with each alternative. This analysis will ultimately reduce the risk of repercussions and their effects.

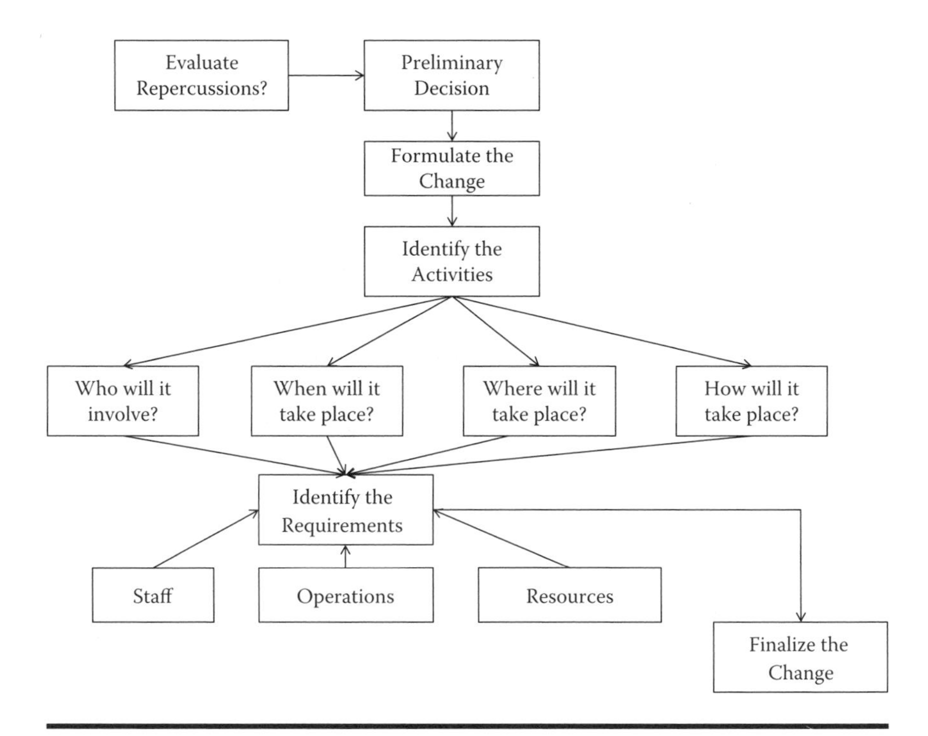

Figure 6.4 Overview the activities.

Alternative and Reprecussion Effects Analysis
(AREA) Note: Failure = Effect of the Repercussion

Process or Product Name:			Prepared by:		Page ___ of ___
Responsible:			FMEA Date (Orig) __________ (Rev) _________		

Process Step	Alternative	Repercussion	Potential Failure Effects	SEV	Potential Causes	OCC	Current Controls	DET	RPN	EOC	Actions Recommended	Resp.	Actions Taken	SEV	OCC	DET	RPN
Identify Process Step (if needed)	What is the Alternative ?	Identify the repercussions (failure points)	What is the effect on the outcome	How Severe is the effect to the project outcome?	What causes or reasons for this Effect ?	How often does a cause occur?	What are the existing controls and procedures that minimze the effect on the outcome?	How well can you detect the cause?			What are the actions for reducing the occurrence of the Cause, or improving detection?	**Who is Responsible for the recommended action?**	What are the actions taken with the recalculated RPN? **Be sure to include completion month/year**				
									0								0
									0								0
									0								0
									0								0
									0								0
									0								0

Figure 6.5 AREA analysis template.

Once complete, formulate the change in terms of its outcome. You should have an excellent idea of what will occur and what the influences will be by initiating the change. If you follow the AREA procedure, there will be control points to minimize any repercussions. Of course, change is now in the concept phase, ready for elements of reality to move it to the design phase.

Now begins the process of identifying the activities in the design phase. Two key elements of this process are

1. Identifying activities and responsibilities (who, what, when, where, and how)
2. Determining the requirements (staff, operations, and resources) for the change

When identifying activities/responsibilities, consider the elements involved with the words *who*, *what*, *when*, *where*, and *how*:

- Who—employees, customers/users, contractors, suppliers/vendors
- What—policies, procedures, processes, products, service offerings, etc.
- How—actions taken, decisions to be made, outcomes to be achieved
- When—time-related issues (implementation time, completion time, etc.)
- Where—geographic, departmental, on-site, off-site, etc.

Use a process map/flowchart procedure, or consider something as simple as a list (Table 6.1). Whatever tool or technique the team decides on or is comfortable with should contain, at a minimum, the information tracked in Table 6.1. The purpose is to be able to visualize the change process and manage it for efficiency and effectiveness. Consider adding additional columns to identify bottleneck areas such as roadblocks, choke

Table 6.1 Change Activity and Requirements List

Change Process Step	*Activity*	*Who Is Involved?*	*What Is Needed?*	*When Will This Occur?*	*Where Will It Occur?*	*How Will It Occur?*	*Requirements*

points, and slowdowns to better facilitate the natural flow. One client uses a checklist, another a detailed process flowchart; it is up to the team to decide. Whatever the choice, the change process should be complete for the design phase.

During this phase, identify the requirements needed to make each process step function well. Focus on qualities rather than just quantity and availability of resources. Rather than just focus on what you need, consider how and who is involved. The complexity is in understanding how these functions and requirements coexist and relate to one another. The task may seem too large, but avoiding such planning results in weeks or months of lost productivity, increased costs, and lost revenue.

During this planning phase, it is critical to communicate with employees. Employees want to know that

- Their work environment remains constant
- Knowledge and experience are relatively static
- The change is for the benefit of the organization, its employees, and its customers

In fact, customers or users want similar information, given their investment in the organization. Sharing the particulars is not required, but reassurances cover a multitude of concerns.

Summary

A critical phase in the change management process now involves the innovation team. Identifying repercussions addresses the negative side of change. Although the execution of the change process could be flawless, repercussions could last for weeks or months. Those who plan and remediate these repercussions will greatly reduce this timeline.

A second feature of this phase is detailing the specifics of the change process. It is not enough to know what will be accomplished, but also who, when, where, and how. Knowing the details of the process is not enough; good planning also requires a needs analysis for resources, support, and internal operations. At this point, the particular change process is now reality and requires one last evaluation.

Discussion Questions

1. Consider some decision you made recently regarding your vacation (holiday) plans. If those traveling with you were not happy with the choice, do you give them at least one alternative?
    a. Did you reach consensus, or was the result left open? Undecided?
    b. Were the alternatives risky in any manner?
    c. What about repercussions and their overall effect?
2. Consider a simple decision process.
    a. Put together a change activity and requirements list.
        i. Does this simplify or make the decision more complex?
    b. Does it help to know who, how, when, where, and what?

V—Validate the Changes

At this stage of ENOVALE, it may be best to re-review the objectives for the proposed change (Figure 6.6). What should that change accomplish? Ultimately, will the change result in

- A return to permanency (consistency)?
- Renewed order and stability?
- A sense of leadership and solid position for the future?
- A set of specified benefits to the organization, including
    - Cost or profit
    - Efficiency or effectiveness
    - Quality, productivity, improved performance
    - Personal traits consistent with the function

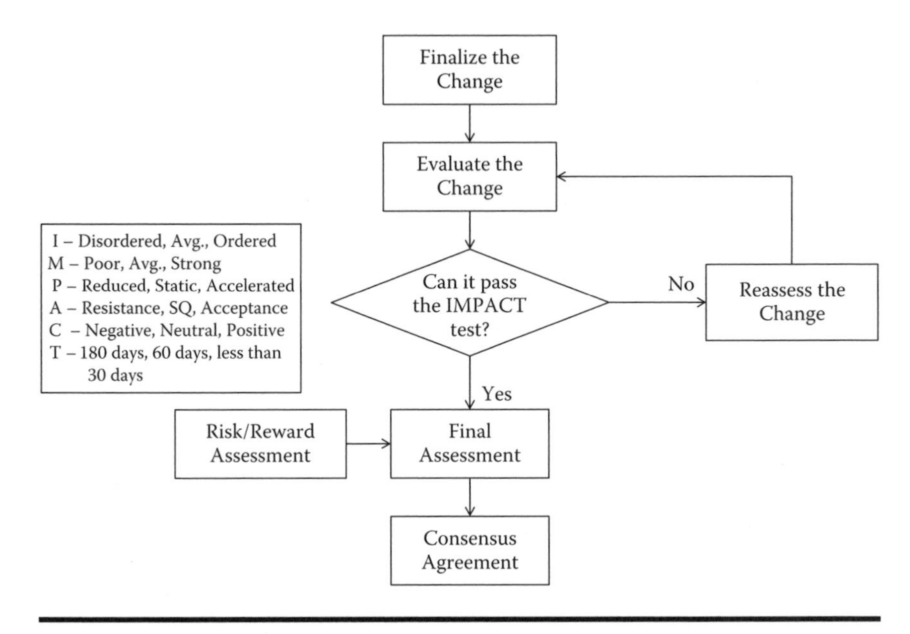

Figure 6.6 V—validate the changes.

If the team and leadership can say that a positive outcome is probable, then there is evidence of innovative change. This step provides "one last look" before implementation. Raise any issues that may have remained dormant. Therefore, prior to deploying, validate all critical issues. Those issues that must precede the implementation of the change include the following:

- Communication provided and received validates the need for the change:
  - A rational explanation of why the change is needed
  - How the change will be managed (if possible, include the specifics)
  - Benefits and positive aspects of the change
  - If a reduction in force (RIF) is to occur, and how it will affect those that remain
- Identify the selection process chosen to eliminate employees. Be sure it does not discriminate.
- The organization remains committed to its values
- Actions to be taken; are they:

  - Justifiable?
  - Minimizing disruptions, displacement, and lost time?
  - Maximizing deployment time?
- Outcomes expected; are they:
  - Reviewed for inconsistencies?
  - Meeting the needs of the organization and the customer/user?
  - Generating the benefits expected?

Use the IMPACT test to evaluate the overall change process (see Table 6.2). The greater the number of positive responses, the better the possibility for innovative change. IMPACT uses a generic time frame for a change management process. Modify the timeliness to meet the requirements of any project. Finally, assess risk and reward from an outcome perspective.

Will the organization receive the benefits it seeks with this change? The final component of validation is approval and consensus. If the team has followed the first four steps, then what follows should function as expected. Unlike the ENOVALE strategies for "new and improved," the organization implements change before the process is complete. The remaining three ENOVALE strategy steps occur during the implementation phase to help keep the project on task and performing as expected.

Table 6.2 IMPACT Statements

IMPACT	*Less than Expected*	*Expected*	*Exceeds Expectations*
Integration	Disordered	Average	Ordered
Managed change	Poor	Average	Strong
Performance	Reduced	Static	Accelerated
Acceptance	Resistance	Status quo	Acceptance
Communications	Negative	Neutral	Positive
Timeliness	Greater than 180 days	30–60 days	Less than 30 days

Summary

Validation is a critical step for change. Unlike previous phases, which highlighted measurement validation, here the team reviews the change process, and if it meets the expectations of the outcome, it is approved. The focus is on the effects of the change, rather than the change itself. The team overviews the planned change process and considers the implications of change. The first implication is that of people, since this is the group most affected by change. Second, how does the change affect daily routines and overall operations? Third, how does the change affect the individual? If the team is comfortable, then use the IMPACT test to review the entire change management process. If all appears as expected, they reach consensus and approve the project.

The main purpose of validation is to identify irregularities in the change management process. It also provides a final review of the rationale for making this decision. Making a decision in an environment of trust, no matter how painful, will increase recovery times. In addition, the results of the process have fewer legal and ethical problems, resulting in less pain and difficulty.

Discussion

Examine a recent change from both the task orientation and human impact perspectives.

- Did the change accomplish its original intention?
- Was the overall change management process successful?
- Were critical requirements superseded due to changes in organizational needs?
- Which elements suffered or benefited the most?

Examine the effectiveness of a change with the IMPACT test.

A—Align for Best Fit

Alignment refers to the process of bringing individuals together to accept and support a unified outcome. For innovative change, this represents acceptance and support of a decision and its consequences. It is not enough to accept the decision; one must also support the results (consequences). This requires that leadership take responsibility for the decision/results and the process of informing/involving employees. For alignment to succeed, the individual must agree to the responsibility of accepting and supporting the decision. Alignment is the process to achieve this goal (Figure 6.7).

There are two critical aspects of alignment:

1. Alignment to the decision:
   a. Once the decision is finalized, then allot a reasonable amount of time for acceptance. Encourage frank and honest discussion, and use evidence to validate the decision.

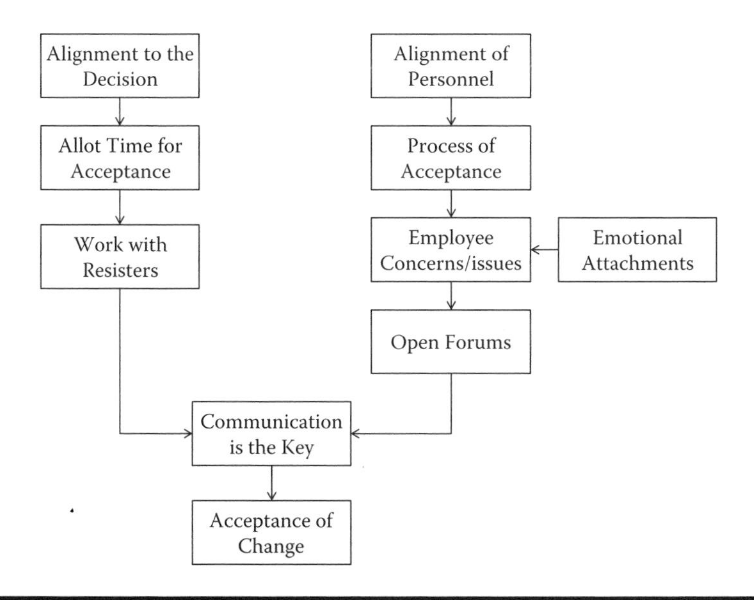

Figure 6.7 A—align for best fit.

   b. Recognize and work with the resisters (to change). They will identify numerous reasons why the change will fail. Listen to their complaints, ask for their suggestions, and use evidence to demonstrate validation of the decision.
   c. The key to eventual acquiesce is communications. Communicate the specifics, especially as the change affects employees.
2. Alignment of personnel to support the decision:
   a. This is the difficult portion of alignment. Employees need to understand a reason for the change and how other alternatives were considered and rejected. Discuss the criteria used to make these decisions. For positive change, this will be simple. For neutral change, there is the possibility of moving this attitude to the positive (or at least the positive side of neutral). For negative change, it may help those remaining to adjust, accept, and cope. Negative change is much like the death of a loved one—there is time needed for grieving, acknowledgment, and finally, acceptance.
   b. In any case, present positive benefits to all.

Aligning to the decision is often a much simpler process than aligning individuals to support the decision. Providing a rational and caring assessment of the situation and the reason for the change is critical. Stress why the decision is the correct one. Discuss the decision process and how the benefits outweighed the negatives. Discuss the method for evaluating alternatives so that individuals understand the reasoning made to accept a final decision. Obviously, if the change is negative, give those affected by the change time to adjust and reset their expectations. Perceptions of the process are critical. It is not just the decision that is being evaluated by all, but the process and the way leadership executes the decision. This is why communication is so critical for any type of change. An effective approach or process will ensure that future changes are implemented with success in a shorter period.

Alignment of personnel to support decisions made should be ongoing within the organization. Prepare employees for the proposed change, as the decision process is ongoing. Consider the following issues:

- Consider the interest and concerns of employees. How will the change affect them both physically and emotionally?
- Reassurance is critical at this stage. This is especially true with negative change, since each individual often feels somehow substandard (not needed). His or her emotions affect his or her behavior and productivity.
- Understand emotional attachments to the existing paradigm (the present way of doing things).
- Be considerate of employees' feelings and emotions. Realize the process will take time, and that management and leadership should be there to reinforce the decision. Making excuses will only make things worse.
- Permit discussions and questions in the context of the decision being made.
- Encourage solutions that would otherwise mitigate the effect of the change.

Summary

Alignment is a critical phase in the overall ENOVALE change management process. Alignment requires management commitment and effective leadership. Implementation will challenge management the first time, but with practice, this will become the norm in future situations. If the intention is to work for positive change, implementation will be a short and simple process. Aligning to the decision and aligning personnel are elements of the alignment process. Both offer challenges and rewards. Organizations proficient in alignment have significantly less lost time, fewer challenges with employees, and a workforce aligned to project success.

Discussion Questions

1. Discuss the strategy used to present a negative outcome. (Consider translating the negative outcome into a hypothetical event.)
   - What would the message contain?
   - How would you deliver the message?
   - How would you explain the decision-making process?
   - How would you ensure that the decision was implemented?
2. Now consider a positive change event. (Again, you could use a hypothetical event.)
   - What would the message contain?
   - How would you deliver the message?
   - How would you explain the decision-making process?
   - How would you ensure that the decision was implemented properly?

L—Lead the Effort

First, change needs to be managed (Figure 6.8). This is a paradigm shift for many who feel that change is merely a decision. The change occurs and then people react to that change. The difficulty in this approach is that change is something that affects individuals, both emotionally and physically. A good change management process within an organization is a key to success. More often than not, change comes with disastrous consequences; most organizations never recover from a major change—they just consolidate or worse, fail. Consider all of the bankruptcies that have occurred. What is the percentage of businesses that have actually succeeded, that have actually grown, after recovering from bankruptcy? The number is quite

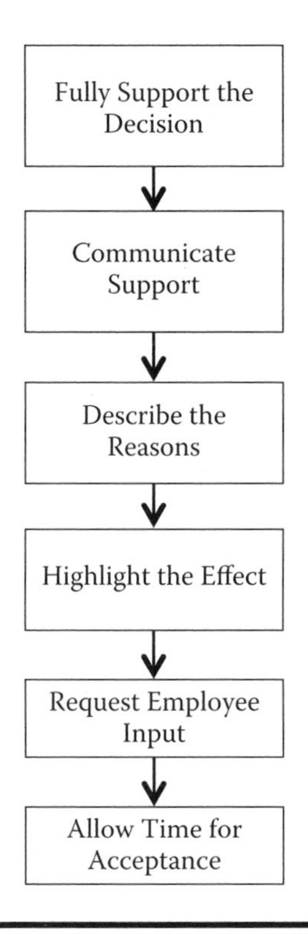

Figure 6.8 Lead the effort process step.

small (approximately 25%) (Dawley, Hoffman, and Brockman, 2003), suggesting that change is far more complex to "turn around." Organizations cannot easily recover from a negative outcome. This is why a managed approach to change can yield significant benefits. In fact, change can be innovative when managed to a positive outcome. "Leaders and organizations can create competitive advantage when they create a culture of high trust and personal connection that unlocks and empowers the untapped capabilities, overcomes the withheld commitment, and dissipates the reluctant distrust so prevalent in other leadership models that lack an authentic interest in the welfare of employees and other stakeholders" (Caldwell and Dixon, 2010, p. 10).

Change is a process, actually quite a natural process, affecting all. In fact, some people have suggested that the theory of evolution is actually the theory of change. That is, nature tries many variations until it reaches one that is better suited or adapted to the environment in which it finds itself. The description in the previous sentence details the true nature of change. Any geneticist reading this book would probably differ from our opinion. However, given the nature of change, it is most productive with the process and a goal.

Effective change needs a leader. The leader must be a unique individual who is both task oriented and possesses human relations skills. That is, the change leader may not be an executive. The ENOVALE process provides a roadmap for change. Change is at best ineffective, or even dangerous, if led in a vacuum. An effective leader provides both the direction and the catalyst for change. Humans respond to change by seeking leaders, during times of change (upheaval), who provide and support viable (although not always the best) solutions. When change occurs and leadership is either ineffective or not engaged, the outcomes are always certain to be negative or neutral at best. The person best qualified to lead change should possess the following traits or skills:

- Excellent communication skills
- Consideration for all employees
- Ability to lead under harsh or compromising circumstances
- A compassionate but deliberate approach
- Able to make decisions based upon evidence and experience
- Identified as a trusted and valued employee (this is the most critical aspect)

Communicating the causes and reasons for change is critical to success. Discuss the expected outcome and how it affects the business, customers, and employees. People adjust best to change when they have some idea of what to expect. If there is time, solicit feedback from customers and employees. In times of change, people must honestly assess their future. Communicating with employees and customers greatly reduces the fear and anxiety that may grip them.

Finally, allow time for acceptance. Negative aspects of change are nearly identical to the loss of a loved one. There is a process (series of steps) that we all go through until we finally accept the fact that what was is no longer. Provide time for adjustment and for those greatly affected by the change—time to pursue new opportunities. Acceptance comes when the individual has fully internalized the change and can move forward with his or her life.

Summary

Leading change is a critical element for success. Not everyone has the skills or desire to lead change. Leading change requires both empathy and persistence. Having a process in which change can occur greatly reduces errors, false starts, and high costs (both physical and emotional). Moderating the impact of change comes with strong communications, probable outcomes, and allowing time for acceptance and personal adjustment.

Discussion Questions

Consider a change that occurred while you were working in an organization. Explain how the change was implemented.

- Was it effective?
- Were communications a key element of the change?
- Was there a leader, and did this person succeed?
- Did employees or customers have time to accept the change?
- Was the change implemented in a systemic manner?
- Could you have improved the process?

E—Embrace the Change

Finally, change affects everyone; no one escapes its reach. Once the leader is named (and the first six ENOVALE steps complete), it is time to rally around the change. Individuals experiencing positive change are better prepared for the next round of change. Implementing change will carry with it challenges that have not been previously addressed. Good planning, in the early stages, will prepare the leader and team to handle consequences and repercussions (Figure 6.9).

As the change unfolds, it is critical to discuss the issues that surface. Change is a flexible process, and modifications may be necessary to adjust to changing business conditions.

Continued communication is critical; solicit feedback on

- Perceptions of and effectiveness with the results (employees and customers)
- Acceptance time and the overall cycle time of change
- General feelings, attitudes, and employee status
- Motivational issues

Leadership and the team have implemented the change. Since the change influences employees and customers, the

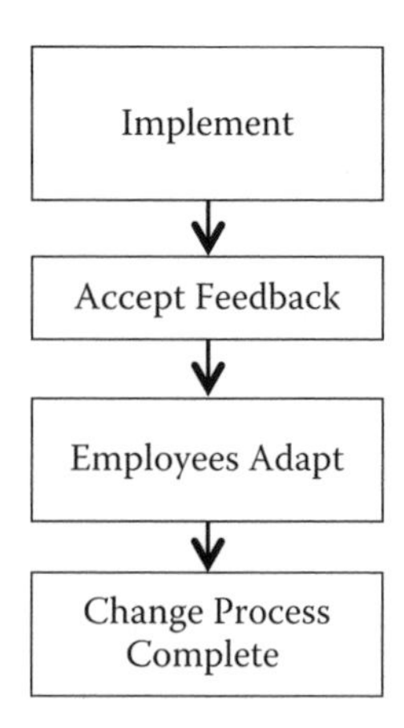

Figure 6.9 E—embrace for change.

focus should remain on them, as they are the center of the change. Trust employees will deal with the new situation and when ready, will feel part of the change. With time, using this approach, employees will be able to embrace change in a more rapid fashion. If the change leads to a positive result, employees will be able to recognize that it is potentially innovative. Allow time for adoption and adaptation. Once the change is accepted, then the individual (including the customer) must adapt to the way of doing business.

For leadership, it may take some time to understand the change to be both positive and innovative. In many instances, the process will be most difficult for this group, as many consider change a decision rather than a process. Leadership is a must, but the leader may come from outside the executive offices.

Finally, change is natural and necessary; organisms are programmed for change. Think about the human body and the frequency of change in cellular structures—aging, muscle strength, eyesight, our immune system—it's all about change!

Summary

In summary, the approach leadership takes is critical for the implementation of the change proposed. By using the ENOVALE process, you will devise a process and management system to use the next time change is required. If planned properly, change can be positive and innovative.

Change can affect employee behaviors and overall productivity. Change has a similar influence on your customers and users. Communicate the change, solicit feedback, and listen to your customers and employees. Describe the reasons for the change, its impact, and overall expected outcome. Truthfulness and honesty go a long way! Prepare the employees and customers for the change, and any disruption time is minimized. Allow time for acceptance of the change and its implications. Change is organic—it's natural.

Discussion Questions

1. Which traits of leadership would be most desirable for change?
2. What type of communications would be beneficial?
   - Before the change occurs?
   - During the change?
   - After the change has occurred?
3. How important is feedback to customers/users and employees?
4. Which is the most preferred method of communicating change?
   - E-mail/electronically
   - Telephone/voice mail
   - Face-to-face encounters
5. Describe change as a living organism.

Chapter 7

ENOVALE for Nonprofit and Agency Organizations

The goal of achieving competitive advantage is not applicable to governmental and nonprofit organizations. These organizations introduce products, processes, services, technologies, and solutions. The difference lies in the fact that these organizations strive to add value and eliminate nonvalue. If a product, process, service, or solution satisfies a new need or expands on an existing need by outperforming previous versions, then the item is innovative.

ENOVALE Process

ENOVALE is a seven-step innovation management process. Steps in the process remain the same, except the link to performance becomes the link to value. This refined sixth step of the ENOVALE process identifies the focus on value (Figure 7.1). Rather than measuring performance, the nonprofit measures value. Values include results such as increased efficiencies,

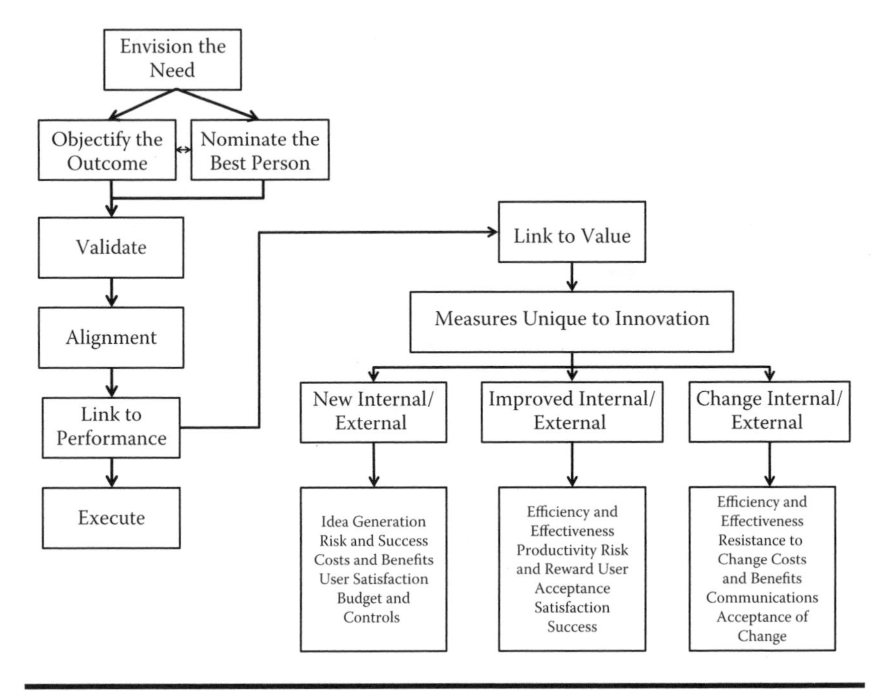

Figure 7.1 ENOVALE process with refined sixth step.

lower costs, increased retention, and satisfied users/constituents. Nonprofits have additional motives for their existence. Some offer general/specific services, professional development, citizen services and responsibility, licensing, etc. Nonprofits play a vital and necessary role in our society. The sixth step of the ENOVALE solutions process easily accommodates the definitional change of performance to value.

ENOVALE Concept of Value

Value, from the concept of ENOVALE, is nothing more than a measure (or evaluation of) expected accomplishment. Some typical internal value measures for nonprofits include the following:

- Efficiencies (time-related metrics, productivity)
- Effectiveness (met or exceeded goals/objectives)

- Budget management (items related to budget and controls)
- Cost reductions (lowest bid acceptance)

Some external measures include the following:

- Accurate and timely information
- Appropriate/applicable solutions
- Service satisfaction

Value is bounded (defined and constrained) by its requirements and expectations. Customers/constituents use the same process to evaluate innovation as they would for a profit organization (see Figure 7.2). The more value that is received, the more expectations are exceeded. Innovation occurs when the value exceeds expectations and satisfies a new or existing need. NASA has continued to develop many new items that have resulted from human space travel. Both the products (technology) they offer and the service provided are

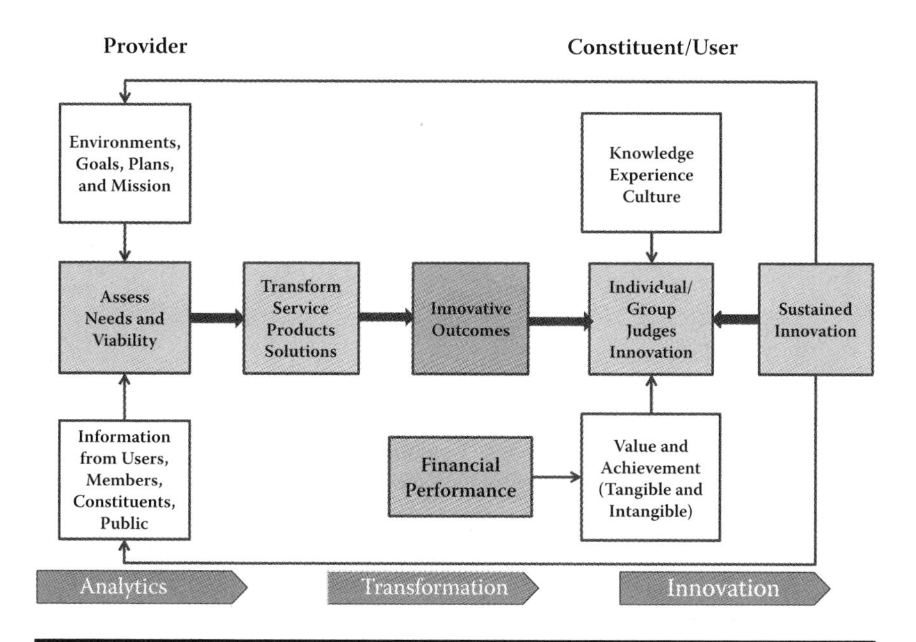

Figure 7.2 Innovation model adjusted for nonprofits.

innovative. They provide real value and drive new products and technology in the for-profit sector.

To judge value, both performance and accomplishment are measured. Innovation brings value to the user (who will judge that the value exceeds what is expected) while meeting or exceeding needs. Nonprofit organizations striving for innovation will use the ENOVALE process (Figure 7.2) to produce innovative outcomes.

Expectations define the human scope of possible accomplishment. Expectations establish the scope of achievement from poor to excellent. Expecting more value from a nonprofit organization is identical to expecting increased performance from a for-profit business. In either case, the human (who judges innovation) shapes his or her expectations based on experiences and encounters. If these encounters are positive, then the opportunity exists for innovation. Negative or neutral experiences do not satisfy users, and therefore cannot be considered innovative. These negative encounters, however, can provide future opportunity for innovation that increases user satisfaction, while adding more value than expected.

For those organizations that do not desire a profit-driven accomplishment, such as a competitive advantage, a substitute objective will replace that particular measure of performance. For a federal agency, the goal may be budget reduction (cost reduction). Value then is a measure of how well the budget cuts meet their expectations (to reduce costs), maintain continuity, and meet all critical objectives. This is the reason for using the term *accomplishment* when describing value.

For nonprofits, revenue generation could be the goal. Here again, value (performance) is the achieved goal. Therefore, innovation truly enables the organization to pursue products, processes, services, or solutions to accomplish its goal in meeting or exceeding customer/user needs. For many nonprofits, raising revenues is a value outcome. The project is innovative if the organization introduced a new campaign, a new approach, or an improved process by meeting new or exceeding existing needs. Innovation defined as new, improved, or change remains 100% applicable; all that changes is the lack of a profit motive.

Concept of Needs

Needs initiate and sustain innovation. This fundamental truth applies to innovation in any form, for any organization. Whether the customer calls himself or herself user, client, citizen, supporter, subscriber, member, patient, or something else, the needs of this person (business, organization, and not-for-profit) drive innovation efforts. Innovation is not limited to high-tech operations; it should be a strategy for all organizations.

The information provided on needs, although definitely slanted toward our initial customers (for-profit businesses), applies to all organizations. The process of obtaining and developing these needs is the same. How these needs are collected, categorized, and classified will vary. Many nonprofits deal with more than one type of customer. Given limited resources, the not-for-profit may deal only with the customer with the loudest voice. Customers (or whatever name you use) require certain needs to be satisfied or they will search for a different way to meet their needs.

Is Innovation a Reality for My Organization?

It will be a paradigm change for many as they begin to incorporate innovation into their organization. The natural reaction is that innovation is for those developing new products or new technology. How will innovation add value to my organization? The outcome or result of innovation is a dramatic improvement (improving an item while meeting or exceeding a need), something new, or a distinctive change. What organization cannot use this outcome? Improvement in innovation is a leap above day-to-day improvements arrived at by such techniques as Lean and Six Sigma. These philosophies teach incremental improvement, which is important to any organization. These techniques do not address needs, but remedy day-to-day problems. Innovation provides the organization with a method to recognize and meet needs while improving service, products, and processes for the organization and end user.

The driving force for innovation may or may not exist in many not-for-profit organizations. Organizations such as professional societies may use innovation to expand their membership. Local and state governments may innovate to cut budgets, increase services, or increase internal efficiencies. If the ultimate goal is to add value, then an innovative approach is worthwhile. Innovation is not difficult, but it requires commitment, resources, and a dedicated workforce for sustained success. ENOVALE provides the process that can operate and produce in any organization as long as there is an end user who desires value and whose needs must be satisfied.

Summary

Needs define the reason for innovation, and value its main deliverable. Whether profit or nonprofit, organizations must offer their customers/users value while meeting or exceeding their needs. For nonprofits, innovation efforts can focus inward to achieve improved efficiencies or outward to achieve additional value. Given Global Targeting's assertion that innovation begins and ends with the individual, innovation in a nonprofit organization excites, aligns, and motivates employees and users alike. If the nonprofit is willing to accept that value is measured and needs are satisfied, then innovation can bring about a dynamic change in the organization.

Appendix A: Measuring Consistency

Consider the graph in Figure A.1, a plot of reliability values (in months). What does it say about consistency?

Look for inconsistency. Since we graph data, it is natural to ask the question: What do the data tell me?

- Anytime a process changes rapidly (and inconsistently), the variation will dramatically increase and it will show this change on a time plot.
- If the data are trending or cycling, this is also an indicator of inconsistency (variability).

Checking for inconsistency:

1. Count runs above and below median. The median is the 50th percentile or the middle value in an odd-numbered data set, with the data ordered from smallest to largest.
   - A run is a series of points on the same side of the median.
   - A run can be any length from 1 point to many points.
   - Too few or too many runs are important signals of changing variation—they indicate something in the process has changed.

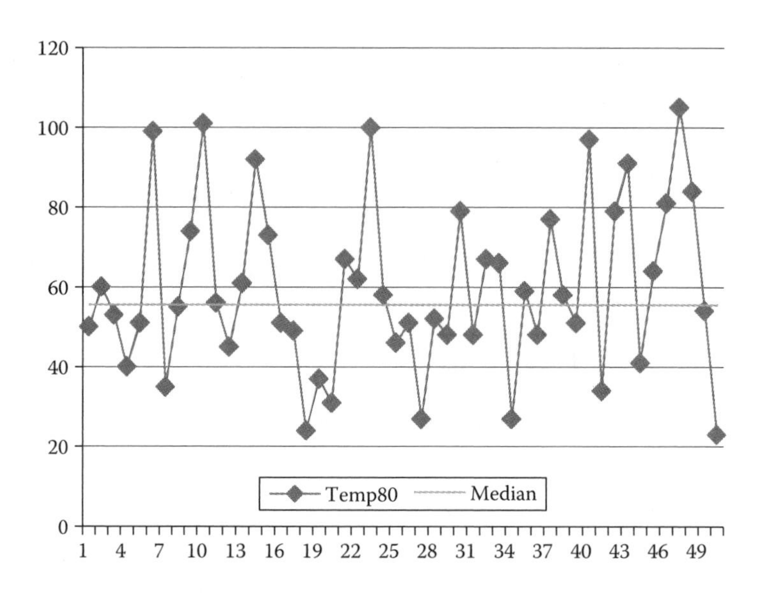

Figure A.1 Plot of performance results.

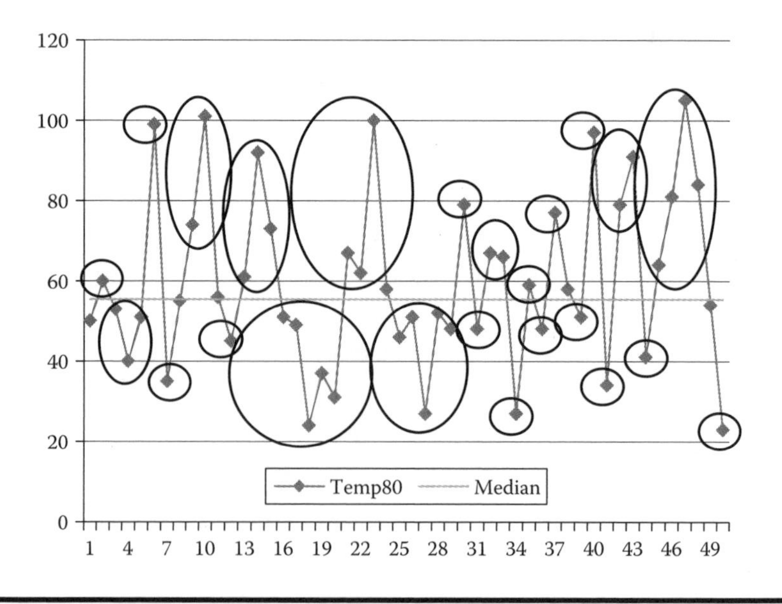

Figure A.2 Run chart analyses (runs identified).

   - Because you often count runs on a time plot, they are also called run charts (Figure A.2).
2. Check unusual occurrences (called special causes).
3. If you see any signals of special causes, try to determine the cause. Work to remove it permanently. If the process is consistent, continue with data analysis.

For the example data, there are no unusual points or inconsistencies; the data are consistent around the median. The high and low values are just part of the process, although their occurrence is rare.

Any line (between two points) that breaks the median breaks or eliminates the run. From Table A.1, determine if the data are unstable (inconsistent). Figure A.3 provides an example of interpreting runs. Figures A.4 and A.5 display patterns, cycles, and trends that all result in a set of inconsistent data. The purpose of this simple tool is to provide a warning when a set of data are inconsistent (not predictable). If the data are inconsistent, it indicates the process is changing.

Table A.1 Runs Table: Examples of Data Inconsistency (Variation)

Number of Data Points Not on Median	*Lower Limit for Number of Runs*	*Upper Limit for Number of Runs*	*Number of Data Points Not on Median*	*Lower Limit for Number of Runs*	*Upper Limit for Number of Runs*
10	3	8	34	12	23
11	3	9	35	13	23
12	3	10	36	13	24
13	4	10	37	13	25
14	4	11	38	14	25
15	4	12	39	14	26
16	5	12	40	15	26
17	5	13	41	16	26
18	6	13	42	16	27
19	6	14	43	17	27
20	6	14	44	17	28
21	7	15	45	17	29
22	7	16	46	17	30
23	8	16	47	18	30
24	8	17	48	18	31
25	9	17	49	19	31
26	9	18	50	19	32
27	9	19	60	24	37
28	10	19	70	28	43
29	10	20	80	33	48
30	11	20	90	37	54
31	11	21	100	42	59
32	11	22	110	46	65
33	11	22	120	51	70

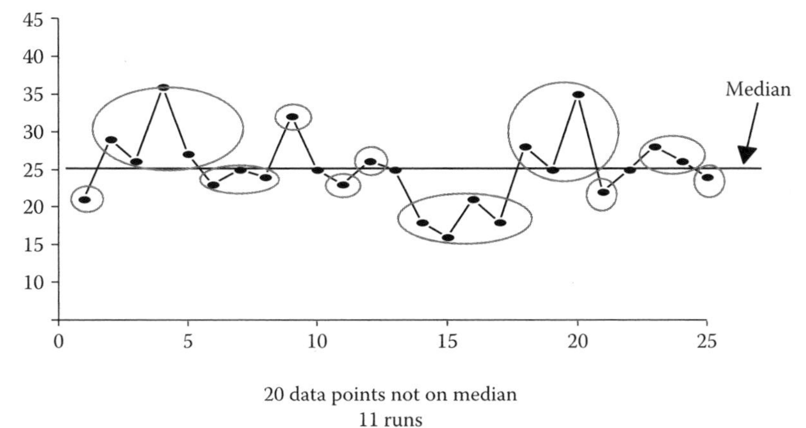

20 data points not on median
11 runs
Note: Points on the median are ignored. They do not add to or interrupt a run.

Figure A.3 Example of interpreting runs.

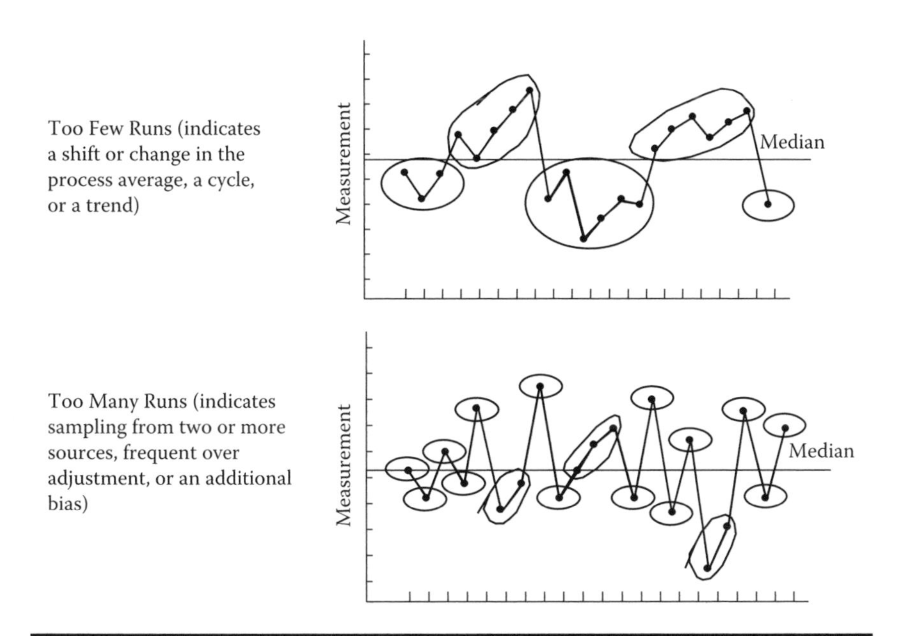

Figure A.4 Examples of inconsistent patterns or cycles.

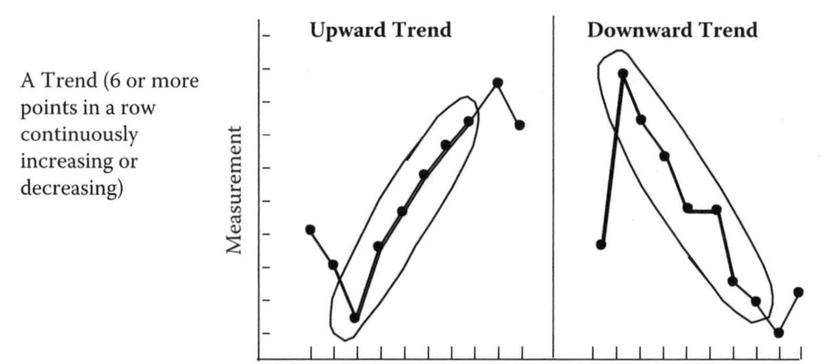
A Trend (6 or more points in a row continuously increasing or decreasing)
Measurement
Upward Trend
Downward Trend
Note: When counting points, omit (ignore) any points that repeat the preceding value. Such points do not add to the length of the run nor break it.

Figure A.5 Example of a trend.

Appendix B: Process Mapping

Process mapping is a helpful tool for those that have a process with multiple steps (Figure B.1). Many find this a useful technique in diagramming a process or a decision. Its primary purpose is to visualize the process and identify potential conflicts, process inconsistency, or conflict (wasteful steps). Lean and Six Sigma methodologies consider process mapping a primary tool in identifying opportunities for improvement. For innovation concerns, process mapping may become critical at the implementation phase. The need to examine alternatives, consequences, and risk make this a useful tool.

This book uses a detailed micro-process map (flowchart) for each step in the ENOVALE® process. If the process or task has numerous steps and decision points, then this tool is highly suggested.

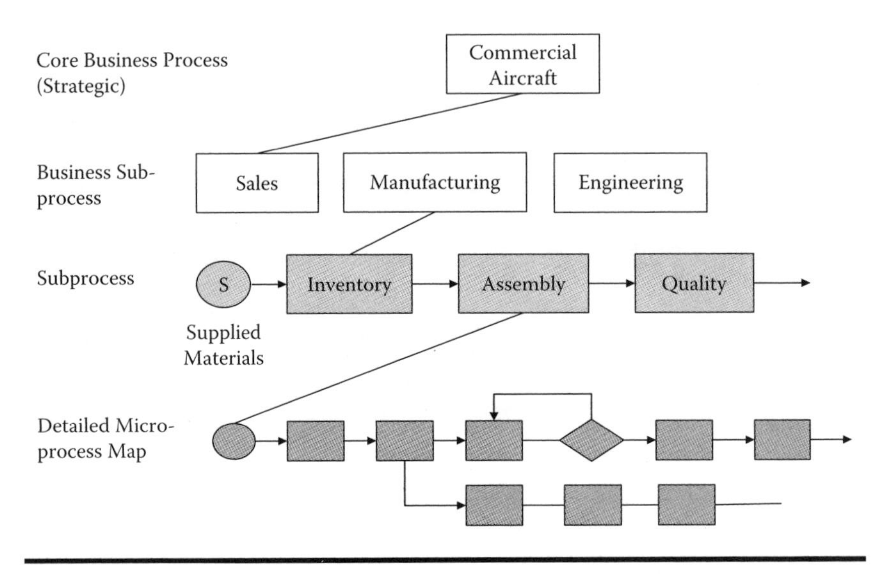

Figure B.1 Hierarchy process map.

Appendix C: Cause-and-Effect (Ishikawa) Diagram

The Ishikawa diagram is an excellent tool for understanding cause-and-effect relationships (Figure C.1). The steps to construct one are as follows:

1. Identify the effect (often the result or its effect).
2. Brainstorm the possible causes.
3. List everyone's causes (no evaluation at this time).
4. Associate each cause with a specific header (category): people, materials, machines, environment, methods, and measurement.
5. Assign priority to the causes in relation to the effect (how strong or weak the relationship).
6. Identify the top three to five causes—and their relationship to the effect.

Note: The cause-and-effect diagram is a great team tool that permits a fair evaluation of all causes without bias.

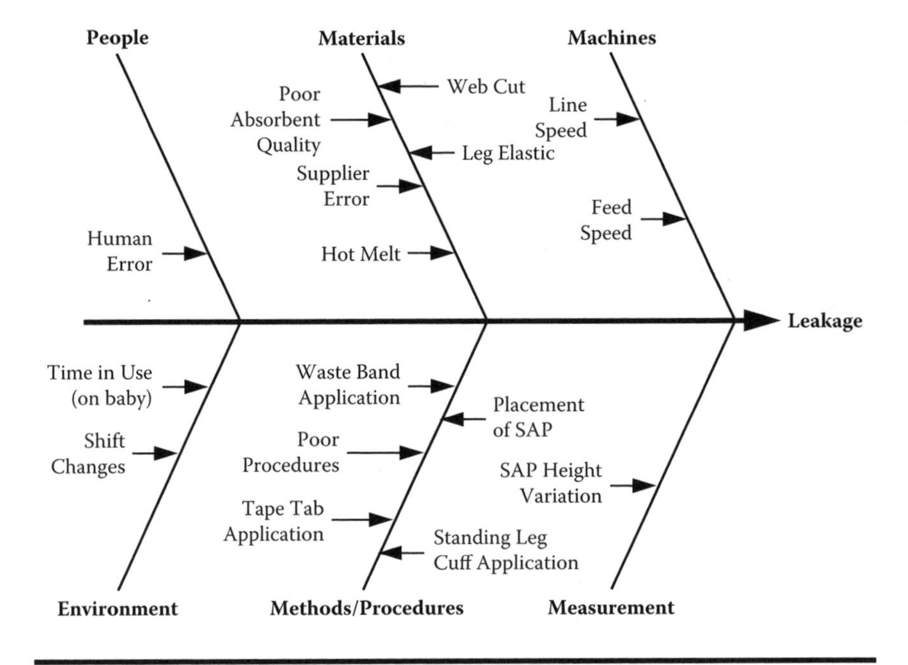

Figure C.1 Cause-and-effect diagram (fishbone chart).

Appendix D: Cause-and-Effect Matrix

This simple matrix enables the innovation team to rank and identify numerous causes. Its purpose is to

- Relate identified causes to critical effects by using customer requirements or needs as a weighting factor
- Emphasize the importance of understanding customer needs and requirements by using a simple quality function deployment (QFD) type matrix
- Help when the number of effects is large
- Score effects with regard to importance to the customer
- Score causes with regard to their relationship to effects

The results are

- A listing of critical causes to evaluate in the FMEA and control FMEA stages
- An ongoing evaluation of the needs and requirements

Steps to create a cause-and-effect matrix are

- Identify needs, requirements, or pair (the interface of need and requirement).

- Order and assign a priority factor to each need, requirement, or pair (usually on a simple scale).
- Evaluate association (correlation) of each cause (process step) to each need, requirement, or pair. Priority scores should contain no more than three levels!
  - 1, 3, and 5 (overall evaluation)
  - 1, 3, and 9 (looking to identify key contributors)

Assignment scores:

- 0 = No correlation.
  - 1 = The cause (process step) only remotely affects the need or customer requirement.
  - 3 = The cause (process step) has a moderate effect on the need or customer requirement.
  - 9 = The cause (process step) has a direct and strong effect on the need or customer requirement.
- Cross-multiply correlation values with priority factors and add across for each input.

Cause-and-Effect Matrix—Focused Approach

An Excel template for the cause-and-effect matrix is available at http://www.globaltargeting.com (Figure D.1).

Phase I

- Place the needs or requirements across the top of the matrix and rank.
- Place the process steps down the side of the matrix.
- Correlate process step to outputs.
- Identify top three to five process steps.

Phase II

Start a new cause-and-effect matrix with the inputs from the top three or four process steps:

- General approach.
- Place the needs or requirements across the top of the matrix and rank.
- Place causes down the side of the matrix, starting with the first process step and moving to the last.
- Assign priority scores and assignment scores.
- Compute correlations and total scores—look for prime cause.

Cause and Effect Matrix

Rating of Importance to Customer (needs) or Project Objective (requirements)																		
			1	2	3	4	5	6	7	8	9	10	11	12	13	14	15	
Choose either a Cause or Process Step			Need/Requirement	Need/Requirement	Need/Requirement	Need/Requirement	Need/Requirement	Need/Requirement	Need/Requirement	Need/Requirement	Need/Requirement	Need/Requirement	Need/Requirement	Need/Requirement	Need/Requirement	Need/Requirement	Need/Requirement	Total
	Process Step	Cause																
1																		0
2																		0
3																		0
4																		0
5																		0
6																		0
7																		0
8																		0
9																		0
10																		0
11																		0
12																		0
13																		0
14																		0
15																		0
16																		0
17																		0
18																		0
19																		0
20																		0
																		0
																		0
Total			0	0	0	0	0	0	0	0	0	0	0	0	0	0	0	

Figure D.1 Cause-and-effect matrix.

Appendix E: Success Modes and Performance Analysis (SMPA)

The SMPA is a tool similar to FMEA, except that it examines the positive aspects of increased performance (achievement). The process begins with the SMPA worksheet (Figure E.1).

The process for completing a SMPA worksheet follows:

1. Identify those elements that affect performance. These elements could be an input, part, policy/procedure, new technology/equipment, etc.
2. How can these elements be improved (that directly affects performance)? Be specific and descriptive. For each element, describe how it could improve performance.
3. What is the effect on performance (consider such items as sustainability, maintainability, and reliability)? The focus is on long-term improvement.
4. What could cause the component, part, or element to affect performance negatively?

Once the worksheet is completed, then use the SMPA template (Figure E.2). An Excel Template is available at http://www.globaltargeting.com.

Use the SMPA rating scale to evaluate those elements that influence or affect performance, the chance that these elements

Success Modes and Performance Analysis (SMPA) Worksheet

Input	Potential Success Mode	Potential Performance	Potential Causes	Current Actions/ Controls
What is the component, part, or element?	In what ways can the component, part, or element improve?	What is the affect on Performance?	What could cause the component, part, or element to affect performance negatively?	What actions are needed for this improvement to be sustained?

Figure E.1 SMPA worksheet.

Success Methods and Performance Analysis
SMPA

Process or Product

Team:

Prepared by:

Page ____ of ____

SMPA Date (Orig) ____________ (Rev) ____________

Process Step	Key Process Input	Potential Success Modes	Potential Performance	P R B	Potential Causes	N E P	Current Actions or Controls	S U S	S P N	Actions Recommended	Resp.	Actions Taken	C O N	O C C	D E T	S P N
Process Step	What is the component, part, or element?	In what ways can the component, part, or element improve?	What is the affect on Performance?	Impact Probability - Rate the chance of continued improvement	What could cause the component, part, or element to affect performance negatively?	How frequently would a negtaive affect occur?	What actions (controls) are needed for this improvement to be sustained?	How well can the improvement sustain increased performance?	Success Priority Number	What are the actions requried for mainataining improved Performance?	Whose Responsible for the recommended actions?	What are the completed actions taken? **Be sure to include completion month/year.**				
									0							0
									0							0
									0							0
									0							0
									0							0
									0							0
									0							0
									0							0
									0							0
									0							0
									0							0
									0							0
									0							0
									0							0
									0							0
									0							0

Figure E.2 SMPA template.

SMPA Scale

Score	Impact Probability	Score	Negative Effect on Performance
10	The SM always or nearly always Impacts the PE	1	The cause will always affect the PE
9	The SM Impacts the PE more than 85% of the time	2	The cause will affect the PE nearly always
8	The SM Impacts the PE more than 75% of the time	3	The cause will affect the PE frequently
7	The SM Impacts the PE more than 60% of the time	4	The cause will affect the PE often
6	The SM Impacts the PE slightly more than 50% of the time	5	The cause will affect the PE more than 50% of the time
5	The SM Impacts the PE about 50% of the time	6	The cause will affect the PE about 50% of the time
4	The SM Impacts the PE less than 50% of the time	7	The cause will affect the PE less than 50% of the time
3	The SM has a small impact on the PE	8	The cause rarely affects the PE
2	The SM has little impact on the PE	9	The cause very rarely affects the PE
1	The SM has no impact on the PE	10	The cause will not affect the PE

Score	Sustainability
10	Sustains PE all of the time
9	Sustains PE nearly all of the time
8	Sustains PE most of the time of the SM
7	Sustains PE about 2/3 of the time
6	Sustains PE more than 50% of the time
5	Sustains PE about 50% of the time
4	Sustains PE slightly less than 50% of the time
3	Rarely sustains PE
2	Very rarely sustains PE
1	Never sustains PE

SM- Success Mode
PE - Performance

Note:
If you cannot sustain performance, then the improvement may not be worthwhile

Figure E.3 SMPA scale.

will perform on a sustaining basis, and causes of problems that would affect performance negatively (Figures E.3 and E.4). Scales are adjustable for the elements chosen. Remember that the occurrence (OCC) scale is an inverse scale, with 10 being the most negative effect. The sustainability scale relates to the ability to sustain improvement over time by controlling negative influences. The purpose of the SMPA is to determine which key process inputs or components can improve and sustain performance.

Process or Product Name:		Improve Wait Time					Prepared by:			Page ___ of ___
Team:							SMPA Date (Orig) ___		(Rev) ___	
Process Step	**Key Process Input**	**Potential Success Modes**	**Potential Performance**	**PRB**	**Potential Causes**	**NEP**	**Current Actions or Controls**	**SUS**	**SPN**	**Actions Recommended**
What is the process step?	What is the component, part, or element?	In what ways can the component, part, or element improve?	What is the effect on Performance?	Impact Probability - Rate the chance of continued improvement	What could cause the component, part, or element to affect performance negatively?	How frequently would a negative effect occur?	What actions (controls) are needed for this improvement to be sustained?	How well can the improvement sustain increased performance?	Success Priority Number	What are the actions requried for mainataining improved Performance?
	Customer's Available Time	Less Wait Time	Judged more efficient	10	Problems with Routing	6	Modify software	5	300	
	Customer's Available Time	Less Wait Time	Judged more efficient	10	Problems with Routing	6	Increase operators	1	60	
	Customer's Available Time	Less Wait Time	Judged more efficient	10	Problems with Routing	6	Increase menu options	10	600	check feasibility of increased menu selection
	Customer's Available Time	Less Wait Time	Judged more efficient	10	Software	8	Purchase or design software	8	640	check available hardware for purchase
	Customer's Available Time	Less Wait Time	Judged more efficient	10	Human interaction	8	Training, follow-up	6	480	

Figure E.4 Sample SMPA decision criteria.

Appendix F: Failure Mode and Effects Analysis (FMEA)

This technique has been available since the late 1960s. It was developed by NASA to examine failure aftermaths and prevent future occurrences. It enables the team to examine situations where an eminent failure can cause problems. Unlike success modes and performance analysis (SMPA), which looks at elements already performing, FMEA examines those elements constantly causing underperformance. The goal is to eliminate or control these defects.

Products are defective because they

- Deviate from the intended condition by the manufacturer
- Are unsafe due to design defects even though they are produced perfectly
- Are incapable of meeting their claimed level of performance
- Are dangerous because they lack adequate warnings and instructions

Design defects:

- Affect the product, service, or technology
- Are built in to the process

- Will greatly reduce future problems if minimized

Technological defects:

- Are due to design, development, resource, and requirement flaws
- Affect performance greatly and are difficult to resolve quickly

Service defects:

- Do not meet the defined criteria for the design and customer
- Are process oriented and may be due to faulty procedures or policies

Types of FMEAs include the following:

- System: Used to analyze systems and subsystems in the early concept and design stages. Focuses on potential failure modes associated with the functions of a system caused by the design.
- Design: Used to analyze product designs before they are released to production. Focuses on product function.
- Process: Used to analyze manufacturing and assembly processes. Focuses on process inputs.

Steps to construct a FMEA include the following:

1. Identify a process that creates a defect (mistake, error, unacceptable product, lack of repeatability, etc.).
2. Identify the major critical elements (or parts) that cause the defect.
3. Complete the FMEA worksheet (Figure F.1).
4. Transfer the information to the FMEA template and complete the template (Figure F.2).

Process Failure Modes and Effects Analysis (FMEA) Worksheet

Critical Element	Potential Failure Mode	Potential Failure Effects	Potential Causes	Current Controls
What is the critical element or part?	**In what ways can this go wrong (fail)?**	What is the consequence on Performance?	What causes or reasons for the loss of performance?	What are the existing controls and procedures (inspection and test) that prevent loss of performance?

Figure F.1 FMEA worksheet.

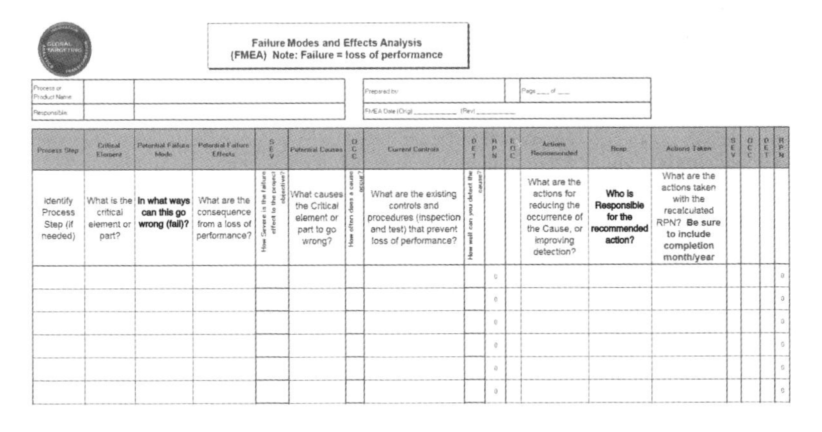

Failure Modes and Effects Analysis (FMEA) Note: Failure = loss of performance

Process or Product Name			Prepared by		Page ___ of ___
Responsible			FMEA Date (Orig) ________ (Rev) ________		

Process Step	Critical Element	Potential Failure Mode	Potential Failure Effects	SEV	Potential Causes	OCC	Current Controls	DET	RPN	EOC	Actions Recommended	Resp.	Actions Taken	SEV	OCC	DET	RPN
Identify Process Step (if needed)	What is the critical element or part?	**In what ways can this go wrong (fail)?**	What are the consequence from a loss of performance?	How Severe is the failure effect to the project objective?	What causes the Critical element or part to go wrong?	How often does a cause occur?	What are the existing controls and procedures (inspection and test) that prevent loss of performance?	How well can you detect the cause?			What are the actions for reducing the occurrence of the Cause, or improving detection?	**Who is Responsible for the recommended action?**	What are the actions taken with the recalculated RPN? **Be sure to include completion month/year**				
									0								0
									0								0
									0								0
									0								0
									0								0
									0								0

Figure F.2 FMEA template.

5. Assign the rating scales (Figure F.3).
6. Apply (calculate) the risk priority number (RPN). An Excel Template is available at http://www.globaltargeting.com. The RPN is a calculated number based on information you provide regarding:
   - The potential failure modes
   - The effects

FMEA Scale
Note FM = cause of failure, FE = loss of performance

Score	Severity	Score	Occurrence
10	The FM always or nearly always impacts the FE	10	The cause of a FM always or nearly always occurs
9	The FM impacts the FE more than 85% of the time	9	The cause of a FM nearly always occurs
8	The FM impacts the FE more than 75% of the time	8	The cause of a FM occurs frequently
7	The FM impacts the FE more than 60% of the time	7	The cause of a FM occurs often
6	The FM impacts the FE slightly more than 50% of the time	6	The cause of a FM occurs slightly more than 50% of the time
5	The FM impacts the FE about 50% of the time	5	The cause of a FM occurs about 50% of the time
4	The FM impacts the FE less than 50% of the time	4	The cause of a FM occurs less than 50% of the time
3	The FM has a small impact on the FE	3	The cause of a FM rarely occurs
2	The FM has little impact on the FE	2	The cause of a FM very rarely occurs
1	The FM has no impact on the FE	1	The cause of a FM never occurs

Score	Detectability
10	Never detect the cause of the FM
9	Very rarely detect the cause of the FM
8	Rarely detect the cause of the FM
7	Detect the cause of the FM about 1/3 of the time
6	Detect the cause of the FM less than 50% of the time
5	Detect the cause of the FM about 50% of the time
4	Detect the cause of the FM slightly more than 50% of the time
3	Frequently detect the cause of thè FM
2	Nearly always detect the cause of the FM
1	Always detect the cause of the FM

Note:
If you can't detect the cause or Failure Mode, the Control is not working!

Figure F.3 FMEA evaluation scale.

- The current ability of the process to detect the failures before reaching the customer

RPN = Severity score * Occurrence score * Detection score = Final numerical value

If the RPN is low, then effective controls exists; if the RPN is high, there is a need for new or better controls.

7. Interpret results, actions, and controls to be implemented.

Appendix G: Control FMEAs

A control failure mode and effects analysis (FMEA) is technique built upon the FMEA. Five basic elements of the control FMEA are

1. List the improve element.
2. List the actions/tasks that will be used to control each improve element.
3. Associate one or more failure modes to each control element.
4. Determine and prevent (contingency and remedial plans, checks and balances) these failure modes.
5. Assign risks based on the severity, occurrence, and detectability of the failure modes to affect the project goal.

Complete the worksheet and template, much like in the FMEA (Figure G.1).

Use the control FMEA to prevent a loss of performance (a defect) and identify specific controls to monitor to prevent this defect. The purpose of the control FMEA is to

1. Ensure that all the necessary elements of the control phase are addressed
2. Serve as a tool for ensuring that the project goal will be achieved and maintained

Control FMEA Worksheet

Process Step	Improve Elements	Control Elements	Failure Modes	Potential Causes	Contingency Plans	Remedial Plans
What is the process Step being Improved?	Which critical elements were selected for Improvement?	What are the actions that will be taken to control each Improve Elements?	In what ways can the Control Element fail?	What causes the Control Element to fail?	What are the plans to prevent the Failure Mode?	What are the checks and balances?

Figure G.1 Control FMEA worksheet.

3. Focus solely on the dynamics of the control (similar to the process FMEA discussed in week 1 training)

An Excel Template is available at http://www.globaltargeting.com (Figure G.2).

To calculate the risk priority number (RPN), using the control FMEA scale (Figure G.3).

The final portion of the document deals with plans to keep the process performing. The innovation teams will create two types of plans. A contingency plan is an alternative set of

Failure Modes and Effects Analysis (FMEA) - Control Process

Innovation Team project #		FMEA Date:	FMEA Revision #

Control Process													
FMEA No.	Process Step	Improve Elements	Control Elements	Failure Modes	Severity	Potential Causes	Occurrence	Contingency Plans	Responsibilities	Remedial Action Plan	Detection	RPN	Audit Item
FMEA No **(Tracking only)**	What is the process Step being Improved?	Which critical elements were selected for Improvement?	What actions will be taken to control each Improvement Element?	In what ways can the Control Element fail?	How severe is the Failure Mode to meeting project objectives?	What causes the Control Element to fail?	How often does the Cause occur?	What are the plans to prevent the Failure Mode or actions to take when failure occurs?	For each Failure Mode, list who, what, when, and how the remediation will occur	What are the checks and balances? **(Document the Remedial Action Plan)**	Will the checks and balances be able to detect non-compliance?	What is the Risk Priority Number?	Yes/No? **(A high RPN requires an Audit Plan)**
												0	
												0	
												0	
												0	

Figure G.2 Control FMEA template.

	Rating			
High 10		Severity	Occurrence	Detection
		Project objective impossible to acheive	Very high and almost inevitable	Cannot detect or detection with very low probability
		Sustained loss of project objective	High repeated failures	Remote or low chance of detection
		Tempory reduction in project objective	Moderate failures	Low detection probability
		Minor impact on project objective	Occassional failures	Moderate detection probability
Low 1		No effect	Failure unlikely	Almost certain detection

Figure G.3 Rating scales for control FMEA.

actions used when the control elements cannot prevent performance from slipping (varies unpredictably due to a known or unknown consequence). A remedial plan is a check and balance against the contingency plan. The control FMEA keeps a high-efficiency process running!

Appendix H: Area and Repercussion Effects Analysis

This tool, similar to a failure mode and effects analysis (FMEA), is useful for examining alternatives and repercussions. Select an alternative, list the repercussions, evaluate their influence on the outcome, and decide if there are means to minimize negative or neutral effects.

To complete the chart, begin with the AREA worksheet (Figure H.1). Transfer the information on the AREA template (Figure H.2). An Excel template is available at http://www.globaltargeting.com.

Evaluate the severity (SEV), occurrence (OCC), and detectability (DET). Use the scales listed below (Figure H.3). Calculate the risk priority number (RPN). Is the alternative worth the risk?

Alternative and Repercussion Effects Analysis (AREA) Worksheet

Critical Element	Potential Failure Mode	Potential Failure Effects	Potential Causes	Current Controls
What is the Alternative?	Identify the repercussions (failure points)?	What is the effect on the outcome?	What are the causes or reasons for the Effect?	What are the existing controls and procedures that minimize the effect on the outcome?

Figure H.1 AREA worksheet.

Alternative and Repercussion Effects Analysis (AREA) Note: Failure = Effect of the Repercussion

Process or Product Name			Prepared by:		Page ___ of ___
Responsible			FMEA Date (Orig) ___________ (Rev) ___________		

Process Step	Alternative	Repercussion	Potential Failure Effects	SEV	Potential Causes	OCC	Current Controls	DET	RPN	EOC	Actions Recommended	Resp.	Actions Taken	SEV	OCC	DET	RPN
Identify Process Step (if needed)	What is the Alternative ?	Identify the repercussions (failure points)	What is the effect on the outcome	How Severe is the effect to the project outcome?	What causes or reasons for this Effect ?	How often does a cause occur?	What are the existing controls and procedures that minimze the effect on the outcome?	How well can you detect the cause?			What are the actions for reducing the occurrence of the Cause, or improving detection?	Whose Responsible for the recommended action?	What are the actions taken with the recalculated RPN? **Be sure to include completion month/year**				
									0								0
									0								0
									0								0
									0								0
									0								0
									0								0

Figure H.2 AREA template.

AREA Scale
Note FM = cause/reason for failure, FE = negative impact on outcome

Score	Severity	Score	Occurrence
10	The FM always or nearly always impacts the FE	10	The cause of a FM always or nearly always occurs
9	The FM impacts the FE more than 85% of the time	9	The cause of a FM nearly always occurs
8	The FM impacts the FE more than 75% of the time	8	The cause of a FM occurs frequently
7	The FM impacts the FE more than 60% of the time	7	The cause of a FM occurs often
6	The FM impacts the FE slightly more than 50% of the time	6	The cause of a FM occurs slightly more than 50% of the time
5	The FM impacts the FE about 50% of the time	5	The cause of a FM occurs about 50% of the time
4	The FM impacts the FE less than 50% of the time	4	The cause of a FM occurs less than 50% of the time
3	The FM has a small impact on the FE	3	The cause of a FM rarely occurs
2	The FM has little impact on the FE	2	The cause of a FM very rarely occurs
1	The FM has no impact on the FE	1	The cause of a FM never occurs

Score	Detectability
10	Never detect the cause of the FM
9	Very rarely detect the cause of the FM
8	Rarely detect the cause of the FM
7	Detect the cause of the FM about 1/3 of the time
6	Detect the cause of the FM less than 50% of the time
5	Detect the cause of the FM about 50% of the time
4	Detect the cause of the FM slightly more than 50% of the time
3	Frequently detect the cause of the FM
2	Nearly always detect the cause of the FM
1	Always detect the cause of the FM

Note:
If you can't detect the cause or Failure Mode, the Control is not working!

Figure H.3 AREA scale.

Appendix I: Check Sheets

Figures I.1 and I.2 show simple and database-driven data collection sheets, respectively.

Product Code	XXX-G3		Lot Num ber	11272013-PWR8018	
Production Line	Machine 3		Inspector	Ralph W.	
Defect Type		Num ber			
Pin Hole					
Low Reservoir Volum e					
High Reservoir Volum e					
Color Match					
Low Shrink Level					
High Shrink Level					
Labeling					
Cd					
Fluid Viscosity					
Other					
Total Count for Shift		68			

Figure I.1 Simple data collection sheet.

1. Time arrived at site: 12:30
2. Time departed site: 17:30
3. TimeSpentDriving 30-45
4. # of devices on job 17 **Basic Package = 5, then add additional devices.**
5. Installer # 1
Installer_Name Mike Krupski
6. Customer #: 19006032
7. Job #: 1
8. Phone Type Wireless
9. Telmar ☐ Salesman ☑
10. Was customer prepared for your arrival? ☑ Yes ☐ No
If not, why? ☐ No Access ☐ Delayed Access ☐ NoPower ☐ Material ☐ No Phone
Other
11. Amount of time lost?
12. Did you receive a job package? ☑ Yes ☐ No
If No, what documents were missing?
13. Did the installation paperwork clearly reflect the work to be performed? ☑ Yes ☐ No

14. Did the customer's expectations match the installation? ☑ Yes ☐ No
15. Were you able to install the job with the equipment specified? ☑ Yes ☐ No
If no, what did you do?
☐ Additional Devices ☐ Move to Wireless ☐ Installed Less Device
Other
16. Did you have the material on the truck to complete the installation? ☑ Yes ☐ No
☐ If no, was the material in stock in the office?
17. How many additional devices, if any, did you sell to this customer?: 0
18. Was this job a: ☐ Take Over ☐ Add On ☐ Prewire ☑ New Installation ☐ Resale ☐ Conversion
If prewire, how many wire drops did you have? 0
19. Was the Sales Rep: ☐ OnSite ☐ Available by phone ☐ Not Accessible ☑ Job OK, no need to contact Rep
20. Today, did you go to the office for:
☐ Material ☐ Paperwork ☐ Meetings
☐ Training ☐ Time Tickets
21. How much time did you spend in the office today?
22. How much time was spent training customer? <15
23. How much time was spent verifying signals? <15

Figure I.2 Database-driven data collection.

References

Baregheh, A., Rowley, J., and Sambrook, S. (2009). Towards a multidisciplinary definition of innovation. *Management Decision*, 47(8), 1323–1339.

Brust, A. (2012, March 1). Big data: Defining its definition. Accessed from http://www.zdnet.com/blog/big-data/big-data-defining-its-definition/1092/23/13.

Caldwell, C., and Dixon, R. (2010). Love, forgiveness, and trust: Critical values of the modern leader. *Journal of Business Ethics*, 93(1), 91–101.

Dahl, A., Lawrence, J., and Pierce, J. (2011). Building an innovation community. *Research—Technology Management*, September-October, 19–27.

Dawley, D., Hoffman, J., and Brockman, E. (2003). Do size and diversification type matter? An examination of post-bankruptcy outcomes. *Journal of Managerial Issues*, 15(4), 413.

Hey, T., Tansley, S., and Tolle, K. (2009). *The fourth paradigm: Data-intensive scientific discovery*. Redmond, WA: Microsoft Research.

McLaughlin, G. (2012, June 25). Why innovation is so often hit or miss. Innovation Management. http://www.innovationmanagement.se/.

McLaughlin, G., and Caraballo, V. (2013). *Chance or choice: Unlocking innovation success*. Boca Raton, FL: Taylor & Francis.

Mish, F. (Ed.). (1990). *Webster's ninth new collegiate dictionary*. Springfield, MA: Merriam-Webster, Inc.

Peters, T.J., and Waterman, R.H. (1982). *In search of excellence: Lessons from America's best-run companies*. New York: Harper & Row.

Index

N

O